The Manager's Guide to Business Writing

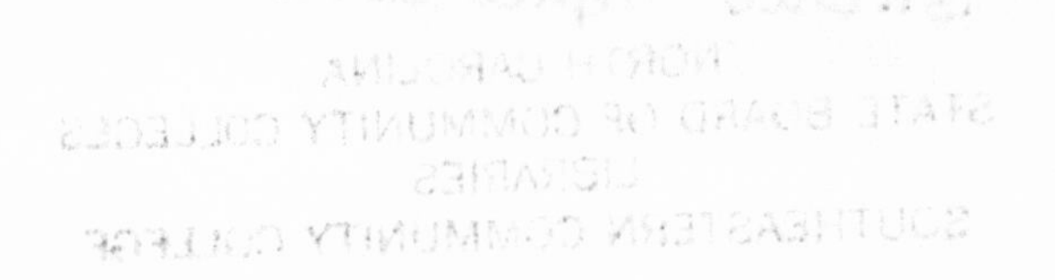

Other titles in the Briefcase Series include:

Motivating Employees by Anne Bruce and James S. Pepitone

Managing Teams by Lawrence Holpp

Effective Coaching by Marshall J. Cook

Performance Management by Robert Bacal

Hiring Great People by Kevin C. Klinvex, Matthew S. O'Connell, and Christopher P. Klinvex

The Manager's Guide to Business Writing

Suzanne D. Sparks

McGraw-Hill
New York San Francisco Washington, D.C. Auckland Bogotá
Caracas Lisbon London Madrid Mexico City Milan
Montreal New Delhi San Juan Singapore
Sydney Tokyo Toronto

McGraw-Hill

A Division of The ***McGraw-Hill*** *Companies*

1 2 3 4 5 6 7 8 9 0 DOC/DOC 9 0 3 2 1 0 9 8

ISBN 0-07-071867-9

Library of Congress Cataloging-in-Publication Data

Sparks, Suzanne D.
The manager's guide to business writing / Suzanne D. Sparks
p. cm.
A Briefcase Book
ISBN 0-07-071867-9
1. Business writing. 2. Communication in management.
HF5718.3.S68 1999
808/.06665 21 98041552

This is a CWL Publishing Enterprises Book, *developed and produced for McGraw-Hill by* CWL Publishing Enterprises, *John A. Woods, President. For more information, contact CWL Publishing Enterprises, 3010 Irvington Way, Madison, WI 53713-3414, www.execpc.com/cwlpubent. Robert Magnan served as editor. For McGraw-Hill, the sponsoring editor was Catherine Schwent, the publisher was Jeffrey Krames, and the production supervisor was Suzanne W. B. Rapcavage.*

Printed and bound by R. R. Donnelley & Sons Company.

Contents

Preface

Writing, like public speaking ranks high on the fear list. Many fear putting their thoughts and feelings into print. Yet, far too many people just throw words out, assuming that others will understand them.

You can't do that if you're a manager because business depends on communication. An effective manager must communicate effectively. But what does that mean? Sometimes it means improving productivity on the job.

Clear communication—the expression of thoughts—improves productivity because it reduces errors. Poor communication harms productivity and efficiency. Common on-the-job errors such as poor project instructions, mediocre memos, lackluster letters, and raincheck (I'll read it later) reports illustrate poor communication at work. Communicating also means making a connection with others. And writing provides a lasting connection.

There's an old proverb that reminds us, "Words fly away; writing remains." What you put down on paper or communicate electronically may not fade away soon; those words may last far longer than you intended—and go much farther.

When you speak, the people around you know the circumstances and take them into account. Maybe you fumble for the right words. Maybe your thoughts and feelings come out in long, wandering tangents. The people around you understand. They know that you don't have the time to organize what you want to say, to choose the right words.

But when you write, you've got the time—or at least people assume you've got the time. They probably expect more of you—better organization, more careful expression. And when you write, you give them the time to read your words, over and over.

Writing can be frightening.

That's why I wrote this book and hopefully why you'll read it. It's important to write well in business. I've tried to focus on your needs, to help you write what you need to write, especially on the job.

Overview

All writing begins with the reader. **Chapter 1** emphasizes that perspective, explaining how you should try to understand the people who will read what you write. From there I move to your reasons for writing. **Chapter 2** discusses the purposes of writing and how we can pursue our purpose through a four-stage process.

The purpose of writing is to express, not to impress. Too many managers forget that basic fact of communication. They use fancy words and long, involved sentences that convey little and make the reader work hard to get at the meaning. We've all laughed at that bumper sticker that proclaims, "Eschew obfuscation." That's the focus of **Chapter 3**. It will show you how to give the most information to your readers and take the least time from them.

Chapter 4 moves into the types of business writing, how you can structure what you write to reach your reader and serve your purpose. It presents the basics of writing letters, memos, and electronic mail.

But writing is not simply a matter of using words. **Chapter 5** shows how we can put more power into what we write by knowing how to use visuals, white space, and headings. The way we frame and complement our words can do a lot to grab attention and guide our readers. **Chapter 6** continues the discussion of attracting and holding the readers. It focuses on two areas that cause the most problems for many people when they write: how to start and how to stop.

Chapter 7 returns to the message with which the book begins: write for your readers. Managers often have to explain their ideas—to their employees, to peers, to supervisors and senior management, to vendors and customers. To do this effectively, you need to know about avoiding assumptions and gaining perspective, about writing descriptions that people can understand and instructions that they can follow.

Chapters 8 and 9 cover two purposes presented in Chapter 2, difficult situations for most managers. In Chapter 8 you'll learn how to deal with tough situations, whether you're announcing bad news or reacting to a challenge, and how to find the positives in negative circumstances. Chapter 9 will help you become more effective at persuading people, whether you're trying to win over a vendor or a customer or sell a proposal to top management.

Chapter 10, the final chapter, shows how to write reports, from the occasional report through activity reports and status or progress reports to the formal report. Maybe you write dozens of reports every year. This chapter will help you gain better results with less effort. Maybe you write only one report a year. Why not make it great? (And this chapter may help you understand how and why you should seize any chance to write a report.)

I end this book with two appendices. **Appendix A** presents my top ten tips for writing well in business. You could actually begin by reading this appendix—if you're in a hurry to improve your writing. If you start there, I believe you'll want to read the entire book. Then, in **Appendix B**, you'll find a quick reference to rules and recommendations for writing effectively. You can find many books about writing. What's the best book? It's whatever book helps you the most to write effectively. I sincerely hope that it's the book you now hold in your hand.

Special Features

The idea behind the books in the Briefcase Series is to give you practical information written in a friendly person-to-person style. The chapters are short, deal with tactical issues, and include lots of examples. They also feature numerous boxes designed to give you different types of specific information. Here's a description of the boxes you'll find in this book.

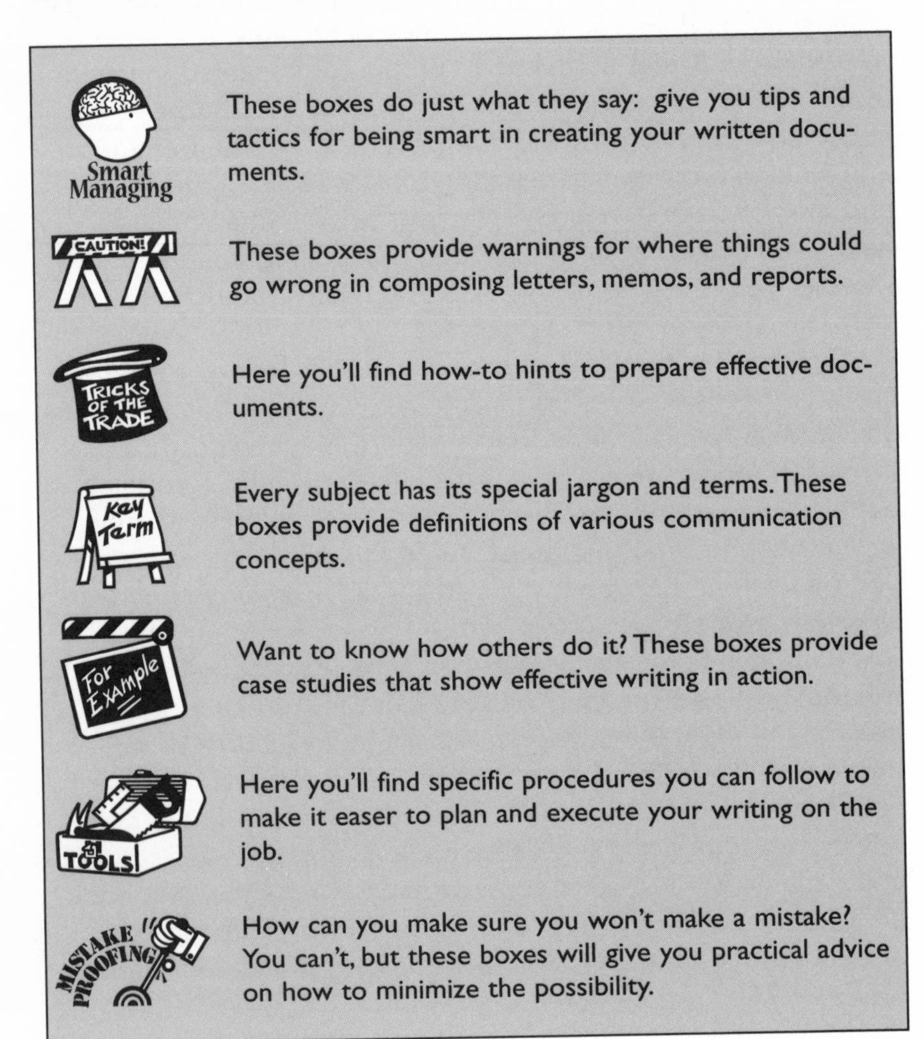

Acknowledgments

I would like to thank Ken Butkus of McGraw-Hill for inviting me to develop books on writing and John Woods of CWL Publishing Enterprises for offering me the opportunity to put into print what I've learned about business writing over the years. I also want to thank Robert Magnan of CWL and Karen Dorman for improving and enhancing my manuscript. Finally, I'd like to thank my colleagues at *communication briefings* newsletter for their valuable input. Finally, I thank you, for reading this book—and for trying to write more effectively.

I dedicate this book to Mom—the great encourager.

Write for Your Readers

Consider...

A set of spotlights containing instructions only in French perplexed and frustrated the English-speaking buyer.

An environmental services brochure that was sent to purchasing managers used such technical language (an "aquifer characterization" and "in situ volatilization to treat the vadose zone") that many confused purchasers chose another source.

A direct mail piece with a pro-life message sent to a pro-choice audience actually caused those people to feel more vehemently opposed to the pro-life position.

A Web page designed for college students tried to arouse a sense of activism in the students; unfortunately, the Web page referred to famous activists like Ralph Nader whom the students didn't recognize.

The dean of a college sent an e-mail to the chairmen of

five departments. Three of the five were women.

Chevy committed a faux pas years ago when it marketed a car called the Nova in Mexico. No va in Spanish means "It doesn't go."

These real-life examples show what can happen when you don't know your audience. Your communication can confuse, anger, or simply fail to connect with the people you want to reach. This chapter focuses on knowing your readers and how you can connect better with them.

Know Your Readers

The first tip to effective writing is to know your audience. The more you know, the more you can tailor or customize your message for an individual or group.

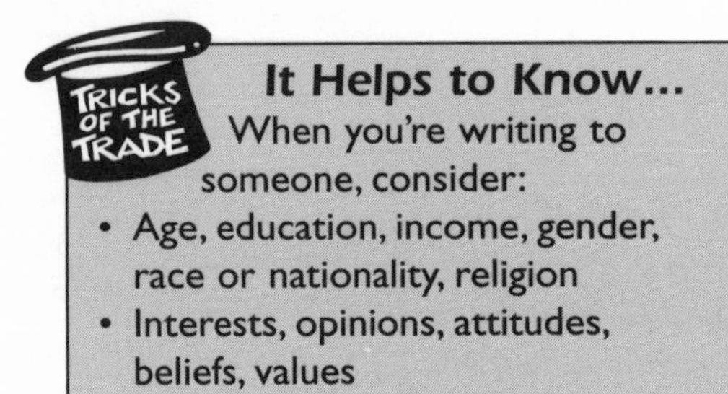

It Helps to Know...

When you're writing to someone, consider:

- Age, education, income, gender, race or nationality, religion
- Interests, opinions, attitudes, beliefs, values
- Reader's knowledge of the topic

First, think of the person or persons you write to most frequently. Visualize your supervisor or your key customer as you write. Try to obtain information such as age, education level, income, and gender.

If you can discover interests, opinions, and values, you can persuade your readers more effectively. You need to know the reader's knowledge of your topic—Is she an expert? Does he know nothing about it? Let's go back to the opening examples.

If the person who wrote the instructions for the spotlights had known the *nationality* of the reader, he could have avoided using the wrong language.

Whoever created the environmental services brochure did not take into account the educational level of the readers or the *readers' knowledge of the topic*, confusing potential purchasers.

The pro-life group ignored the *values* of the people who would read its message.

The person who designed the Web page to arouse activism in college students ignored an important demographic, *age*: the students were too young to remember or care about Ralph Nader.

The dean should have considered *gender* and addressed his e-mail to "chairs" or "chairpersons."

Chevy's marketers ignored the *nationality* of Spanish speakers, who would interpret the car's name as a major negative factor.

You can see that if the writers of these pieces had known their audiences, they could have avoided serious blunders.

What If You Don't Know Your Readers?

The scenario: You have fifteen minutes to write a memo and you don't know much about the manager you're addressing. Here are some quick tips.

> Key Term
>
> **Audience types** Basic categories of readers, according to their knowledge of the subject and their interest: layperson, expert, executive, user, complex, and mixed.

In most cases you just need to spend a few minutes determining which of the following categories most closely fits your reader. Then you can easily adjust your writing.

It's helpful to evaluate whether your reader is a *layperson*, an *expert*, an *executive*, a *user*, or a *complex* type. Here are some guidelines to help you categorize your readers, with some "Dos and Don'ts" and a few examples.

Layperson

A layperson has little expertise in a subject matter and usually no particular motivation to read what you write. So

> Key Term
>
> **Layperson** Someone with little expertise in a subject and usually no particular motivation.

to be effective, you must motivate or attract your reader; starting with a benefit helps. A layperson is not knowledgeable, so you must adjust your tone, style, and vocabulary.

Do: Find a way to attract attention.

Don't: Bore your reader with detail.

For example... If you're writing to employees (laypersons) about various health care plans, find an interesting fact or a reason (benefit) for them to read your first paragraph, like how they can receive 100% coverage for dependents. If you're writing for people who use computers but do not know any software program well, you might attract attention by using an easy-to-understand analogy. You might also present one of the benefits of using a particular software program, like the grammar- and style-checking feature of a word processing program.

Expert

An expert cares about process and detail. An expert who is a chemist, for example, would want to know how to reproduce your results by using all the procedures you followed. Give experts the specifics. The same detail would scare or bore the layperson.

Do: Focus on procedure or process.

Don't: Only give bottom-line data.

For example... If you write to an expert in health care benefits, spell out the details of the policy. The expert will understand and appreciate the specifics. If you're writing about computer software for programmers, you'll want to go into particulars about how you developed a particular program.

Expert Someone with considerable knowledge about the subject and great interest in details.

Executive

An executive audience wants bottom-line information. Detailed descriptions that work

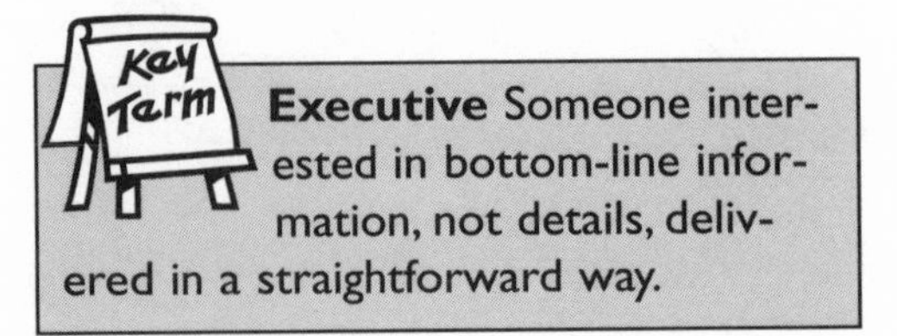

Executive Someone interested in bottom-line information, not details, delivered in a straightforward way.

for experts would not work with this audience. Use straightforward language and tone. Give a benefit and the critical information first.

Do: Get to the point immediately.

Don't: Explain in detail.

For example... Give the executive audience a summary of the medical benefits package in one paragraph or less. Then proceed with other important points. The manager in charge of selling the software product isn't interested in how it works, but in how she will sell it.

User

The user must carry out your instructions. For example, users of a software package must read your documentation in order to do their job. These people don't care how you wrote the software; they want to know how to make it work.

Do: Realize that this person might not know as much as you do.

Don't: Be too brief.

> Key Term
>
> **User** Someone who wants or needs to know how to make something work. Any other information might be superfluous.

For example... The *user* in our health care plan example would need to follow the complicated medical policy. Help the user by explaining clearly how to use each policy. The person who must use the software and understand how he can make it work needs the basics and in sufficient detail.

Writers make a common mistake with user audiences: they overestimate the readers. This error seems to be particularly true in technical matters. In one instance, an employee was trying to use a new software program, but the manual didn't help. It began with the command to type in a password after the prompt. Unfortunately, the employee didn't know how to turn the computer on or to find a prompt, so he was unable to use the software. The writer simply had neglected to start at the beginning, to provide the basics.

Complex

You must write to fit your reader, to establish a connection that will make your writing more effective. This is especially difficult when the reader might be a *complex* audience, a combination of styles. Here are a few examples.

- The person who serves as your supervisor may be a *layperson/executive*, a manager with no particular expertise in your specific field. You must motivate him to read your work. Use benefits to catch his attention and a bottom-line style to keep his interest.
- You might report to an *expert/executive*, an engineer who has worked her way up in the company to become CEO. An executive summary followed by a detailed explanation will work for this CEO.
- A communication manager who still writes and edits newsletters is an *expert/user*. Tell the expert/user how the process works and how she can personally implement it.
- An employee using the Internet could be a *layperson/user*. This person needs motivation and information. With no particular expertise, he may have difficulty accessing e-mail messages through the Internet. Give the layperson/user the necessary information in a way that motivates him to use it.

Sometimes you may write for a mixed audience, meaning that your readers comprise all four types. When writing a company newsletter, for example, you must address laypersons, experts, executives, and users as well as complex types. In this case, you must write for the "lowest common denominator," the layperson.

Dealing with the Differences

Now, let's look at an example. Imagine that you're writing a series of letters to promote your company's newsletter about baseball, *Buzz around the Bases*, and your promotional campaign includes a free copy of a booklet titled *The Story behind Major League Baseball Contracts*. In preparing the pitch letters, you need to appeal to a fan, an agent, an owner, and a player—four very different types of readers. Notice how you write to deal with the differences among your readers!

Layperson (baseball fan)

> Dear Kate:
>
> How do you get to watch your favorite baseball player? How does an athlete make it from the amateur ranks into the big leagues?
>
> An agent, acting as go-between for a team owner and a player, negotiates contracts based on salary caps or limits. When you see your favorite major league baseball player, you may not be aware of the behind-the-scenes discussions among these agents and owners to contract with valuable players.
>
> When negotiating, agents must consider the player's compatibility with a team, length of contract, and available monies. The result you see may be a star player.

Note the simple vocabulary and informal tone. The fan may not care about contract negotiations, so you use a benefit (watching a favorite player) to attract attention. Then you describe in simple terms how an agent negotiates.

Expert (Agent)

> Dear Pete:
>
> To negotiate a major league contract for your new client, you will need to take the following factors into account:
>
> - current rules on salary cap, including how much of the signing bonus counts against the cap,
> - whether players are plentiful or in short supply, and
> - the team's needs versus your player's skill.
>
> Once you narrow the number of teams based on your client's geographic preference, you'll need to obtain comparable salary data among players of similar skills, age, and performance. Once you locate a team on your preference list with the needs that match your client's skills, you'll need to determine whether the team has available money under the cap. If so, begin to negotiate. If not, move to the next team on your preference list.

Note the emphasis on the explanation of the process, with more details, and using terms familiar to Pete. The agent would want to know how—what process to use when negotiating a contract for a player.

Executive (owner)

> Dear Marge:
>
> As an owner, you must assemble the right combination of players and decide what type of packages to offer based on your budget and the income you desire. Depending on your motivation, determine the balance you'd like between your desire to win and the amount of money you want to make.

Note the emphasis on the bottom line and the direct approach. This executive needs to know how to quickly and effectively negotiate for selected players. She's not necessarily concerned about the complete process because she relies on an agent to negotiate for her, so you don't tell her every step the agent must take. This letter is concise, direct, and informative.

User (player)

> Dear Lenny:
>
> How do you negotiate a major league contract? It's relatively easy, but many players make crucial errors.
>
> Select an agent who will represent you well. Decide which cities you would like to live in and how long you would be willing to stay there.
>
> Also, determine the range of salary you would accept. Discuss with your agent the importance you place on a winning team and on your chemistry with other players.

Note the step-by-step approach to telling the player what to do to get the best contract. Lenny needs to use the information you're providing to work with agents and owners to join the team of his choice and make the salary he desires. He must live with the results of the negotiation. Just because he knows baseball doesn't make him an expert in contract negotiations. Spell it out for Lenny in a simple, benefit-oriented way.

Isn't it amazing how you can deal with one topic in several very different ways, depending on your readers? Be sensitive to your particular audience and the response will reflect your

Men and Women

Don't approach men and women in the same manner, because they generally view the world differently says Deborah Tannen, a sociolinguist and author of *You Just Don't Understand: Women and Men in Conversation* and *Talking From 9 to 5: Women and Men in the Workplace*.

Women tend to approach the world as individuals in a network of interpersonal connections. For them, the aim of communication is to create and maintain relationships, to get and give support, and to reach consensus.

Men, on the other hand, usually look on the world as a hierarchy in which it counts to achieve high status and to preserve independence. For men, communication is part of the struggle to gain and keep the upper hand and to challenge others.

effort. If you addressed our executive, Marge, as if she were a player like Lenny or if you wrote to Pete, the agent, as if he were simply a fan like Kate, they probably wouldn't respond as you'd like. Make every effort to adjust your vocabulary, tone, and approach for each type of reader.

Check for Readability

Writing for your readers includes knowing their knowledge level and their interest. Those two key factors vary greatly, depending on whether your readers are laypersons, experts, executives, users, complex, or mixed.

There's another key factor—reading ability. That's not just a matter of literacy. It also involves attention span, the environment (time and distractions), and comfort level. That's why you want to be sensitive to the reading level of your writing, to minimize what experts call fog.

Many managers seem to feel that they should use big words and long sentences to impress their readers. Unfortunately,

Fog Linguistic obscurity, anything in writing that makes the message less clear.

their efforts just tend to fog up their communication and make their writing less effective.

How can you avoid this problem? By testing your writing to determine the *fog index*—then editing it to bring it up or (generally) down to a reading level more appropriate to your readers.

Here's how to determine the Gunning-Mueller Fog Index™, as presented in *How to Take the Fog out of Writing* by Robert Gunning and Douglas Mueller:

1. Select a sample of your writing that consists of 100 words.
2. Divide the number of words in the sample by the number of sentences to get the average sentence length.
3. Count the number of words with three or more syllables in your sample. Don't include proper nouns (names), compound nouns (such as "briefcase" or "bookkeeper"), or verb forms that have three syllables because of a suffix (e.g., "created" or entering" or "advises").
4. Divide the number of long words by the number of words in your sample to get the percentage of long words.
5. Add the average sentence length (from Step 2) and the percentage of long words (from Step 4), dropping the percentage sign. Multiply the sum by .4 to find your fog index.

That index represents the number of years of education needed to understand the writing easily. An index of 7, for example, would mean that the writing is appropriate for a reader with seven years of school, while 12 would be the level of a high school graduate and 16 would be the level of a college graduate.

Let's check the fog index for the following writing sample:

> Knowing your audience might be the most critical factor in effective writing. The more you know, the easier it is to tailor your message for an individual or group. Sometimes all you know is that your audience has a short attention span.
>
> I once wrote a report for eighteen-year-old readers. I refer-

> enced a classic musical, My Fair Lady; unfortunately, this group was too young and had no idea what I was talking about. I lost credibility and the attention of my audience. If you know the demographics of your audience, you can dramatically improve your ability to communicate.

This writing sample conveniently has 100 words in seven sentences, which gives us an average sentence length of 14.3 words. It contains 16 words of three or more syllables; divide that number by the 100 total words and we get 16%. Then, we add 14.3 and 16 and multiply the result, 30.3, by .4. Our answer—12—shows that this sample would be appropriate for readers with at least a high school education.

Most newspapers in the United States are written at a twelve-year-old reading level, a fog index of 7 or so. That's generally a safe index when you don't know the level of your specific readers. If you write too high or too low, your readers may find your writing either difficult to understand or insulting.

But how do you account for situational factors that impede understanding? After all, you can't expect your readers to concentrate on your words if they're in a hurry or trying to do several things at once. If you want to make your writing easier to read without insulting your readers, the best bet is to shorten your sentences (and your paragraphs), but not necessarily to find shorter words. (We'll consider word choice in Chapter 3.)

Now that you know your readers and the best level at which to write for them, how do you get them to read what you write? That's crucial no matter what you're writing. Whether you're developing a pitch letter for your latest product or sending out a memo or e-mail, you want the recipi-

CAUTION!

Fog Alert!

Don't use the fog index on everything you write. Just use it from time to time or when you're concerned about the level of your writing.

Also, remember that the fog index is just a rough guide, since the length of a word or a sentence is in itself no guarantee of ease or difficulty. Plus, punctuation can do a lot to help or hurt the clarity of your writing.

ents of your golden words to actually invest the time and energy to read them.

Writing from the Reader's Perspective

Perhaps the most effective way to get people to read your writing is by taking the reader's perspective. Focus on the benefits for him or her. Why should that person read your e-mail, memo, letter, or report? How will your document benefit the reader?

There's a simple way to work from a "you" perspective. When preparing to write any form of communication—e-mail, memo, letter, report, or whatever—just put yourself in your readers' shoes and ask the question, "Why should I care about what you're telling me?" That should help you focus on your readers from their perspective.

It's easy to take the reader for granted. In fact, we've been doing so in this chapter, using the term "reader" as if we could assume that all those people out there are necessarily going to read what you write. You've got to motivate them to read your words. You've got to hit them immediately with the benefits for them. You've got to answer that old question on their minds, "What's in it for me?" Otherwise, you're less likely to connect with them—and your masterpiece may just get dumped into the vertical file or the recycle bin.

Which of the following two paragraphs would make a better opener to a letter?

Sample A:

This is to announce that as of June 18 Bagin Technology will begin manufacturing computerized controls for power lawn mowers in order to enter an expanding market that will allow the company to take advantage of its leading position in the electronic controls market.

Sample B:

As one of our loyal customers, you should be among the first to know our big news: as of June 18 Bagin Technology will begin manufacturing computerized controls for power lawn mowers.

> You've helped make us the leader in the world of electronic controls and we'd like to show our appreciation for your trust by expanding to serve your needs for electronic controls.

If you received two letters and read those two opening paragraphs, which letter would you continue reading? Where is the reader in Sample A? The information is all from the perspective of the company. Sample B appeals to the reader from the first line through the last.

Build Reader Rapport

Catching the attention of the reader is crucial. But you've also got to hold it long enough to get your message through. What should you do for the reader with a short attention span? What about the reader who's confused or upset or even hostile? And how do you deal with the growing number of people who have just too little time to do too many things?

Try the following tips:

Establish common ground. Begin with something you share. Bond with the reader. If you're writing to persuade your CEO to add an on-site day care center, for example, don't start by listing all the things you need to accomplish. Instead, start by reminding her of your common interest in making the company a better place to work. You can do this even more effectively by writing from her perspective, emphasizing reduced tardiness, better morale, higher productivity, and a lower attrition rate. In other words, don't start with your dream; start with her reality.

Agree (at least partially) with readers if you know their position. Acknowledge the validity of the readers' positions and recognize their objections to your positions. Reduce any adversarial distance. For example, beginning with "I understand why you're reluctant to set up a day care center and I appreciate your concerns about the expenses" is certainly better than opening with "We really want a day care center and we'll do whatever it takes to set one up."

Overcome your audience's objections. Eliminate objections or constraints one by one by providing evidence. If you know the CEO won't want the insurance implications of an on-site day care center, suggest that an outside company come in to run the center and assume liability.

Use short paragraphs to hold interest. Particularly for the audience with a short attention span, keep your evidence succinct and clear. Focus on the reader's perspective. Don't become repetitive just to reinforce your position. State your evidence, dispel any objections, and sum up your position.

Conclude with an action statement. Answer the question, "OK, now what?" When summing up your arguments for a day

TRICKS OF THE TRADE

Consumer Complaints

When responding in writing to a consumer complaint, try these five steps:

1. **Recite the facts.** Repeat them exactly as the consumer outlined them. Cite specific contacts with you or your organization. This will establish that you understand the complaint and intend to deal with the issue.
2. **Empathize.** Say, "How badly you must have felt when you counted on us but our product/service failed you." Don't just sympathize. Empathize. The consumer must know that you understand and that you feel the same.
3. **Put the problem in its place.** Make certain the consumer knows that the problem is an exception. Cite your record, your reputation, the number of satisfied customers. Why? So that the consumer will feel confident that you can solve the problem.
4. **Cite a specific remedial action.** Don't just offer promises about someone looking into the problem. Give the names of people who will work on it and dates when they'll take action.
5. **Reinforce empathy and future contact.** Show your empathy. Repeat the date or time you'll get back to the consumer and invite him or her to contact you again until the problem is solved. Then send a goodwill gift as an apology and a sign of your sincere commitment to resolving the problem.

Source: *communications briefings*, reprinted with permission.

care center, suggest a date to resolve the issue. No imperatives, no ultimatums—just a specific time frame to discuss your plan.

Come to Terms with Gender

Unfortunately, many managers work hard at communicating, but undermine their efforts with language choices, sometimes quite innocently, even unconsciously. We're all aware of the issue of "politically correct" language. What can we do?

First, we should not characterize the issue as "politically correct." That's a disparaging term that shows limited understanding of the issue.

Second, we should acknowledge that we will never make it through life without offending somebody with some word that we use. Our paranoia can only hurt our communication.

Third, we should understand that this issue is fundamentally a matter of sensitivity, of respect, of not making people feel excluded or bad.

The purpose of communication is to connect and convey. Avoid anything that gets in the way of that purpose. Focus on what matters. Words that call attention to unimportant things distract from your message.

It's impossible to avoid offending somebody, but it's easy to minimize the chances of offending. Try these commonsense guidelines:

- Avoid unnecessary mention of physical appearances, race, marital status, or other characteristics.
- Focus on the positive. For example, "physically challenged" makes more sense than "handicapped." (After all, you hired that person because of what she or he could do.)
- Avoid using gender-specific words (such as those listed later in this section).
- Avoid gender modifiers (e.g., "female engineer" or "male nurse"). If gender specificity is necessary, pair the modifiers ("female engineer" and "male engineer"). The key

here is equality: "men and women" or "ladies and gentlemen" or "males and females" or "girls and boys." (If you wouldn't say "man doctor" or "gentleman lawyer" don't say "woman doctor" or "lady lawyer.")

- Refer to women and men in the same way. Equality again. Refer to them all by their first names or their last names or "Mr." and "Ms."

To what extent should we use gender-neutral language? Because English does not have neutral pronouns, we're caught between such forms as "she or he" and "he/she," which may be awkward, and the grammatically incorrect plural "they." It's generally best to use the plural or to alternate between "he" and "she" forms. (See box for examples.)

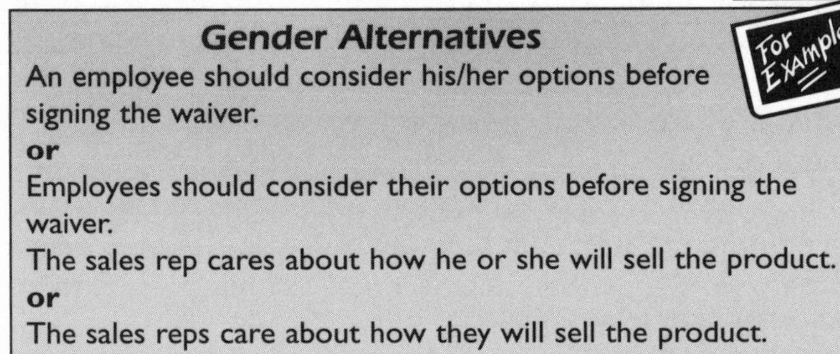

Gender Alternatives

An employee should consider his/her options before signing the waiver.

or

Employees should consider their options before signing the waiver.

The sales rep cares about how he or she will sell the product.

or

The sales reps care about how they will sell the product.

If a sales rep makes a big sale, congratulate him. If a rep lands a new account, give her a bonus.

Here are a few suggestions for avoiding unnecessary gender references.

Gender Sensitive	Gender Neutral
mailman	mail carrier
congressman	congressperson
chairman	chairperson
spokesman	spokesperson
layman	layperson
repairman	repairperson, mechanic
fireman	firefighter

salesman	salesperson, sales rep
policeman	police officer
deliveryman	delivery person
manpower	workforce
mankind	humankind, humanity, society
manmade	manufactured, artificial, synthetic
man-hours	work hours
foreman	supervisor

The bottom line—it doesn't make sense to use words that make people feel excluded or bad or uncomfortable—especially when it takes just a moment of thought to choose more appropriate words. Sure, it's awkward to be using new expressions, but our ancestors somehow adjusted to using "you" instead of "thee" and "thou." We've all had to learn at least a hundred new terms in the past few years just to use our computer systems. If you did that to get along better with machines, what small sacrifices should we endure to get along better with people?

Manager's Checklist for Chapter 1

- ❑ Find out as much as possible about your reader.
- ❑ Take the reader's perspective.
- ❑ When in doubt about your reader's reading level, it's safest to aim at a fog level of 7 or so, a twelve-year-old reading level.
- ❑ Identify your reader—layperson, expert, executive, user, or complex—and shape your message to that person's knowledge and interests.
- ❑ Remember: all readers want to know what's in it for them.
- ❑ Be attentive to the impact of your word choices on other people.

From Purpose Through Process

Before you begin writing anything, decide on your purpose. Do you want to inform your reader? Are you trying to persuade your reader? Do you need to provide instructions for something? Are you writing to record an activity? Research proves that we write more effectively when we have a single purpose rather than several.

Remember your purpose as you write. I once had a professor who wrote "So what?" at the end of some of my papers. If you ask yourself the "So what?" question and you cannot answer it, your reader may experience the same problem.

Know Your Single Purpose

Business writing usually accomplishes one of the following tasks:

- Inform
- Persuade
- Instruct
- Record/document

Here are six examples. If the writers kept a single purpose in mind, you should easily determine that purpose.

Example #1

> Dear Ms. Spagnolia:
>
> We provide the most cost-effective solution in the business. We have the expertise, the personnel, and the resources to design a state-of-the-art fitness center for you. We can increase your productivity and decrease absenteeism by providing this on-site fitness center at a competitive price.
>
> Consider carefully the following benefits of choosing our firm: increased productivity, higher employee morale, decreased absenteeism, and the knowledge that you have selected the premier provider of fitness services.
>
> When you consider how you can have the best company design a fitness center that will provide so many benefits, I'm sure that you'll contact us immediately to begin improving your work environment and your bottom line.

This letter persuades us because it presents the competitive advantages of the company and the benefits for us.

Example #2

> TO: Great Valley Personnel
> FROM: Quality Improvement Team
> SUBJECT: Parking Lot Lights
> DATE: June 20, 1998
>
> Additional lights will be installed on the perimeter of the parking lot this summer. There will be a total of eleven new lights, three on each side lot and five lights along the back lot.
>
> These lights should create a safer environment. If you have any questions, call any member of the quality team. Enjoy!

This memo informs us, because it tells us about something with no expectation that we should act on that information.

Example #3

> TO: Human resources
> FROM: Data entry manager
> DATE: March 7, 1998
> RE: Dorothy Davis file

> On March 6, 1998, Bill Block met with Dorothy Davis to discuss her order entry errors.
>
> After evaluating Dorothy's errors, Bill found no particular pattern to her incidents. He suggested having an experienced customer service representative sit with her and actively monitor her customer service skills.
>
> Dorothy is considered a good customer service representative who achieves the 55% availability ratio. Bill found Dorothy receptive to his suggestions; she hopes to reduce her order entry errors.

This memo records, or documents, Dorothy's problem and Bill's conversation with her.

Example #4

> Welcome to your Mr. Espresso Machine! In order to use this product properly, you must follow the above-mentioned warnings. If you follow the directions, you will enjoy a delicious cup of espresso or cappuccino.
>
> Have a stainless steel container to froth milk. Make sure your machine is plugged in. Place the appropriate amount of espresso in the dispenser, as shown in Figure 1. Be careful not to touch the machine while brewing. Froth milk by inserting the plastic cable into the stainless steel container. Enjoy with your favorite topping.

This excerpt from an owner's manual is meant to instruct the owner on the use of the product.

Example #5

> TO: Fellow Employees
> FROM: Rowan Executive Offices
> RE: Business Strategy Update
> DATE: September 3, 1998
>
> As you know, these times continue to be challenging for the environmental market and our organization. Looking ahead, we'd like to share with you the principles and business strategies that will shape our future, along with the current actions underway to address our present market conditions.

Many of you have expressed concern over the prospects for your own long-term future at Rowan. On the one hand, you see a market continuing to endure the impact of federal budgetary uncertainties. Yet on the other hand, the industry is undergoing an exciting and dynamic shift from regulatory to economic drivers that challenge our traditional market perceptions.

Our well-established strengths of environmental quality and safety remain the foundation of our sustainable development initiative. Our shared task is to grow these traditional attributes. While we refine Rowan's business direction, we will address the implications of our current market conditions by closely aligning our resources to the level of our current project backlog.

Let us assure you that we will work diligently to finalize the new organizational structure by year-end and will keep you apprised of our progress. Thanks for your role in our continued success.

This memo informs employees of management's direction and lets them know where they stand as employees of Rowan.

Example #6

Conference Center Hotels & Resorts
105 College Road East
Princeton, NJ 08540

June 22, 1998

Mr. Alan Dinning, President
SCC Associates
696 Girard Avenue
Aurora, OH 44202

Dear Alan:

As the President of the Conference Center Hotels & Resorts, I am writing to let you know about some important results of our organization that could affect your business with us.

During the past two years, our Princeton Conference Center with its 291 guest rooms has had an exceptional increase in bookings, revenues, and profitability. Our revenue performance per guest room has placed us among the top ten performers among all suburban hotels in the United States, according to Lodging Hospitality's annual statistics for the lodging industry. Our performance and our managerial leadership have made the Conference Center the recipient of numerous awards throughout the years. These awards include the Pinnacle Award, the Paragon Award, and the AAA Four Diamond Award.

We hope to secure your future business with us. We have enjoyed dealing with you in the past and would look forward to exceeding your expectations in the future. We would like to continue the tradition of our award-winning performance results for SCC Associates.

We will call you shortly to discuss your upcoming conference needs. If you need anything in the meantime, please call us. Thank you for your continued business.

Sincerely,

Katrina Alberici

Katrina Alberici, President

This letter attempts to persuade SCC Associates to book future business with the Conference Center because of the results it has achieved and the awards it has received.

Did you determine the purpose of each example? It's not always easy, because sometimes a piece of writing will inform in order to persuade, instruct, or record. Examples 1 and 6, for instance, provide information as a means of persuading. In example 5, the concluding paragraph is definitely more persuasive than informative. We can more easily determine the purpose if we focus less on what the writer is doing and more on why.

As I've already noted, we can be more effective in our writing when we have a single purpose. That may mean, though, that you use the opening paragraph or two in a letter to inform, then follow with a paragraph to persuade. Or you might begin with several informative paragraphs about a product before getting to your central purpose, which is explaining how to use that product and how to take advantage of the features you've detailed.

In general, however, you should limit each piece of writing to a single purpose. This is particularly true of letters, memos, and e-mails: your results will be better if you focus on one purpose. Once you've determined your purpose, you can begin writing.

Writing Purposes

Here are some of the purposes of business writing.

Informative writing: monthly reports, trip summaries, benefit announcements, human resources information, and meeting arrangements.

Persuasive writing: proposals, requests for time off, requests for additional personnel, client correspondence, brochures, and direct mail pieces.

Instructive writing: tutorials, help screens, user manuals, and instructions.

Recording writing: personnel reviews, minutes from meetings, client contact sheets, and time sheets.

Writing in Four Stages

An efficient way to write effectively is in four stages:

- Plan.
- Write.
- Revise.
- Edit.

Sure, you can write things in just one stage (use that kind of writing for notes to yourself, as a memory aid). If you want to express something to somebody else, write in four stages for the best results.

Planning

Planning saves time later in the writing process. If you clearly outline your purpose and content, you'll spend less time revising and editing later. Just as a good presentation requires a plan or outline, so does a well-written report, letter, memo, or e-mail. The

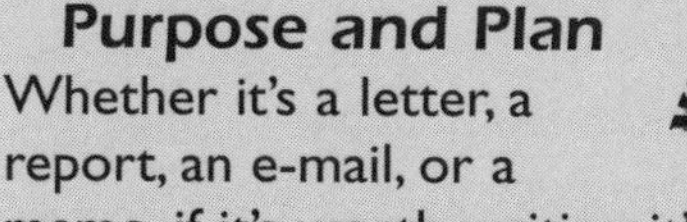

Purpose and Plan

Whether it's a letter, a report, an e-mail, or a memo, if it's worth writing, it's worth writing well. Don't try to save time and effort by just writing off the top of your head.

Start with a purpose. Why are you writing? Then decide what steps you need to take to achieve your objective and the order of those steps. That's your strategy.

thought you put into pre-writing saves you time in rewriting.

Plan how you want to achieve your purpose. Think strategically and organize your thoughts. An outline can help you do this.

An outline? That word probably brings back memories of English classes, of trying to remember whether to use Roman numerals or Arabic, capital letter or lowercase. Maybe you once learned all the rules of outlining, but if you're like many of us you simply left them all behind. After all, do you really need all those rules to make an outline?

No. In fact, some experts recommend other ways to organize, such as with mind maps, which are like mental flowcharts. For those who prefer the traditional outline format, some word-processing software contains features that simplify that task.

But what matters is that you think about what you want to do and that you establish the best order for doing it. Take the first of the sample letters that started this chapter. Here's how the writer might have planned the strategy for persuading the reader to choose that company.

- **Competitive advantages**
  - most cost-effective solution
  - expertise
  - personnel
  - resources
  - competitive price
  - premier provider of fitness services
- **Benefits of fitness center**
  - increase productivity
  - decrease absenteeism
  - boost employee morale
  - confidence in choosing best provider
- **Conclusion**
  - review competitive advantages
  - review benefits
  - encourage to contact

That's a simple outline, but it orders the essential steps for persuading the reader to choose this particular company. You may notice that the writer didn't follow the outline rigorously; some of the benefits were presented in the paragraph detailing the competitive advantages. That's OK, because the outline served its two purposes: it helped the writer keep track of the points to make and it established an order.

You don't need a complex structure with letters and numerals. All you need is a way to plan your strategy.

You might start by simply jotting down ideas on a sheet of paper or sticky notes. Don't worry about organizing at first; just get your thoughts down on paper.

Then, when you've jotted down all the ideas that come to mind, decide on a basic means of organizing them that best suits your purpose. There are several ways to order the points you want to make:

- Chronologically—most often best for instructive writing, but also good if your purpose is to record events.
- Increasing order of importance—especially good for persuading or for expert readers, particularly if readers may not agree with you.
- Decreasing order of importance—often best for informing or for bottom-line readers, when you want to get the vital information up front.

As you're organizing your points, others may come to mind. Jot them down and integrate them. You may also think of certain choice phrases that you'll want to use in the writing stage. Jot them down, too. But don't try to think about writing yet. Focus on coming up with all the points you'll want to make and on putting them in the best order.

Pull together all your thoughts and information into your outline. You may not use everything you've written down, but you'll have it all available so you don't leave anything out unintentionally. In our example above, the writer could have inserted a few facts and figures under "expertise," "personnel," and "resources." Maybe he or she did so, then decided not to use them, so as not to weigh the letter down. Gather everything in the planning stage; you can decide what to leave out when you write and revise.

Writing

Now, take your outline and sit down with your pen and paper or your computer. Are you ready to write?

Not yet! First, take a few moments to create a mental picture of your reader. If you don't know the reader, try to imagine what she or he is like. What makes that person tick? If you're writing to a group of people, put some faces on that group. Think of them as individuals. What differences are there among them? What do they have in common?

Then, when you've got a mental picture of your reader(s), just start a conversation. Use your outline as your guide and write naturally. Use words that come easily to mind. If you can't think of the "right" word, put down the best you've got and move along.

This approach helps your writing come to life and conveys your presence. That makes your words more effective. Don't worry about style. Style will emerge the more you write this way. At this point you should concentrate on putting your thoughts into words. I discuss language in greater detail in Chapter 3, but I can make several basic recommendations here.

Use simple language. Don't use big words when small words will do the job. What's the difference, for example, between "use" and "utilize" or "implement"? If there's no real difference, why use bigger words? Write to express, not to impress.

Be specific. Avoid vague words that can be misunderstood. Don't say "office equipment" if you mean a personal computer.

Use the active voice whenever possible. Instead of "This change is to be made," write "We'll make this change" (it's shorter and it provides more information).

Keep your sentences relatively short. Longer sentences discourage many readers. One recommendation is that your sentences should average seventeen words. That doesn't mean, of course, that seventeen words is the ideal length. To communicate effectively, you should vary the length of your sentences.

Make paragraphs no more than six lines long. With longer paragraphs, many readers will read the first and last few lines and skim or simply skip everything in between. Tight writing invites the reader to continue. Think about breaking up big chunks of text with headlines or lists.

When you've finished writing, set your work aside, if you can, to build some distance. At least take a few minutes to do something else, if only to walk around briefly get a drink, or use the bathroom. When you return to your writing, it should be easier to go on to the revising and editing stages.

Writer's Block

If you have trouble starting, here's a suggestion.

Begin with what comes easiest to you. Maybe it's the ending or some part in the middle. If you've got a good outline, you don't need to worry about writing in order from start to finish. So go back to your outline. Read your points one by one until you find a point that invites you to write.

Revising

Think of this stage of the writing process as quality assurance at a macro level. You need to make sure that the results of your writing achieve the intended purpose(s) as effectively as possible.

Try to read your writing from the perspective of your reader. You might even want to sit in a different chair and to read your words aloud. You'll find out if your copy is conversational and natural. Also, mistakes are more likely to stand out. If you used a computer, you should print out a copy and read your words from the paper, as your reader will be reading them.

It's Written—Now What?

Look at the major pieces of text. Do they all belong? Are they in the most effective order? Do you use paragraphs or sentences that don't add anything to your message, or any words or phrases you don't need? Are you using any long words when shorter words would express the same thought just as well?

If you received the piece, would you read it? Is it attractive? Does it invite the reader?

Be tough on yourself. Be critical. Above all, don't hesitate to make changes: that's the purpose of the revision stage. You may need to move a paragraph or two, drop some text, and/or add something, perhaps to elaborate on one of your points.

If you find that you're making a lot of major revisions, that may mean you need to spend more time and thought in the planning stage, to improve your outlines (especially if you're adding a lot to your text at this stage).

When you have all the pieces you want where you want them, it's time to move on to the final stage, to run your work through quality assurance at the micro level.

Editing

When you edit your writing, you're checking it over for accuracy, grammar, spelling, and punctuation. That's tough, because most managers aren't language perfectionists and because it's always difficult to review your own work. Sure, we catch typos easily enough, but what about those errors each of us tends to make, such as confusing "their" and "they're" and "there" or not knowing where to place an apostrophe? How likely are we to catch those mistakes?

You may want to have someone else edit your writing (after you've done your best). If that's not an option, you can try the old trick of reading the paragraphs in reverse order. Reading your work from finish to start may disrupt the flow of your words enough for you to catch some errors.

Mistakes Your Computer Won't Catch

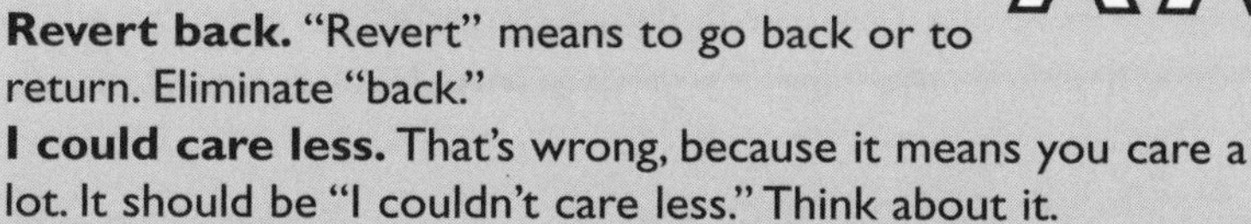

Revert back. "Revert" means to go back or to return. Eliminate "back."

I could care less. That's wrong, because it means you care a lot. It should be "I couldn't care less." Think about it.

New innovation. An innovation is something new, so drop the adjective.

The principle reason. The adjective should be "principal," meaning the most important.

8 p.m. tonight. Because "p.m." means after noon and "a.m." means before noon, we don't need to use qualifiers such as "tonight" or "in the morning" when we're using those Latin abbreviations.

This will effect my job status. The verb should be "affect," which means to influence or to produce an effect upon. The verb "effect" means to bring about.

A final word of advice here: don't trust your computer. Those tools that check grammar and spelling are very helpful (in fact, I'm using them as I work on this book). But they're limited, because human language is simply too complex for machines (at least for now).

Five-Step Revision and Edit

You can do a very thorough check of your writing in five steps. You may not need this process for a memo or an e-mail, but try it for a long report. It moves from the most general level to the most specific.

Three Major Grammar Errors

Between you and I. The phrase should be "between you and me" because "me" is the object of the proposition "between" and must be in the objective case.

Here's several ways to… The contraction doesn't work, because what follows is plural (it's "Here are several," not "Here is several").

The company issued it's report. "It's" stands for "it is" or "it has." Because we want to indicate that the report belongs to the company, we use the possessive "its" (no apostrophe).

1. **Overall purpose and content**—the "So what?" question. Read only for purpose and content. Decide if your content makes sense and your purpose seems clear.
2. **Overall structure and organization.** Part by part, look at how your report is organized. Does each part fit where it is? Do the parts all together make sense in this order?
3. **Paragraphs.** Look at each paragraph to see if each sentence belongs there. Is each paragraph complete? Does it achieve its purpose?
4. **Sentences.** Does each sentence make sense? Is each sentence really a sentence? (Effective writing can include sentence fragments, because they're a natural part of normal language use. But you should choose to use fragments, not just allow them to happen, which can make your writing seem sloppy.)
5. **Your personal errors.** Read especially closely for those mistakes that you tend to make. You may even want to keep a checklist around your desk to make sure you won't miss any of your personal favorites.

Help for Your Grammar

Grammar seems so difficult for those of us who haven't thought of Miss Thistlebottom since sixth grade. Fortunately, the Internet provides terrific sources. You can try www.wordwizard.com to ask questions or receive help with your grammar.

Help is also as close as your phone. *Copy Editor*, a newsletter for copy editors, lists sixty-seven grammar hotlines in the United States and one in Canada. I can't list them all here, but I can recommend calling a

MISTAKE PROOFING

The Ten Least Wanted

Here are the ten words most frequently misspelled by people in business:

accommodate	incompatible
a lot	occasion
commitment	separate
develop	supersede
embarrass	inadvertent

university professor in Nashville who operates a grammar tips hotline (615/353-3349) or calling the University Writer's Helpline in Philadelphia (215/204-5612) for answers to your general writing questions. Of course, I also provide some grammar tips in Appendix A.

Manager's Checklist for Chapter 2

- ❑ Know your single purpose or objective: to inform, persuade, instruct, or record.
- ❑ Plan and revise your writing. Don't just write and edit.
- ❑ Use the style, tone, and vocabulary most appropriate for your purpose.
- ❑ Consider a five-step revision for a more thorough check of your writing.
- ❑ Keep track of your personal grammatical problems and be particularly attentive to those problems when you edit your writing.

The Power of Words to Express, Not Impress

Which of the following two paragraphs would you rather read?

1. In order to ensure that the process of recording calls by work order number is properly operational, it is necessary to purge the work order assignments and related two-digit account code assignments for each telephone extension and update the system to include only those work order numbers for active client projects.
2. Use only active client work order numbers (and their two-digit account codes) to record calls. Purge inactive work order numbers and their account codes.

Sometimes writers try to impress, not express. When we use words to impress our audience, we may lose them by writing at a high fog index. When we write to express, we focus our attention on the reader, not on ourselves, and on making our writing readable.

Make It Clear, Economical, and Straightforward

What does "readable" mean? How do you make your writing more readable?

Readability is the result of many factors. The three most important are clarity, economy, and straightforwardness.

Clear writing conveys meaning without ambiguity. To write well in business, you must make sure that the reader at least understands you.

Economical writing uses no more words than necessary. That's one of the distinguishing marks of clear and forceful writing. To write well in business, you must make sure not to waste the reader's time and energy.

Straightforward writing puts words in a natural expected order, such as placing the subject close to the verb for easy understanding. To write well in business, you must make it easy for the reader to know what you mean.

Clarity

If your message has more than one meaning, it's not clear. Don't use long words where short ones will do; it makes your writing dense and difficult to understand. Use precise words and phrasings to make your writing clear. Make sure the words you choose have the right meaning and don't allow for misinterpretation. For example, don't use the vague term *health organization* when you mean the *American Red Cross.*

Economy

When it comes to words, more is not usually better. When you use too many words, you may lose the reader—or at least make the reader lose interest. Take the following paragraph, for example.

> For the purposes of this policy, "sexual harassment" may be defined as unsolicited nonreciprocal behavior by an employee who is in a position to control or affect another employee's job or who uses the power or authority of that position to cause that employee to submit to sexual activity or to fear that he or she would be punished for refusal to submit to such activity. Sexual harassment also includes any employee conduct reasonably interfering with another employee's work performance by creating an intimidating, hostile, or offensive working environment.

It's easy to get lost in that jungle of words. How effective will that policy be when the clarification is so dense? The following sentence provides another example of uneconomical writing.

> Parking in the lot adjacent to the building will be restricted by space allocation designation for workers' vehicles and the four outermost spaces will be reserved for supervisors of the construction crew so employees should make other arrangements for parking during that time frame and consider implementation of vehicular co-transportation.

What should you do instead? First, determine what information in the sentence you want the reader to understand. Then, eliminate any unnecessary phrases or redundancies. Try one of these revisions:

> Consider carpooling, because parking next to the building is primarily for workers.
>
> Because we are designating parking spaces for construction workers, we suggest that you carpool.

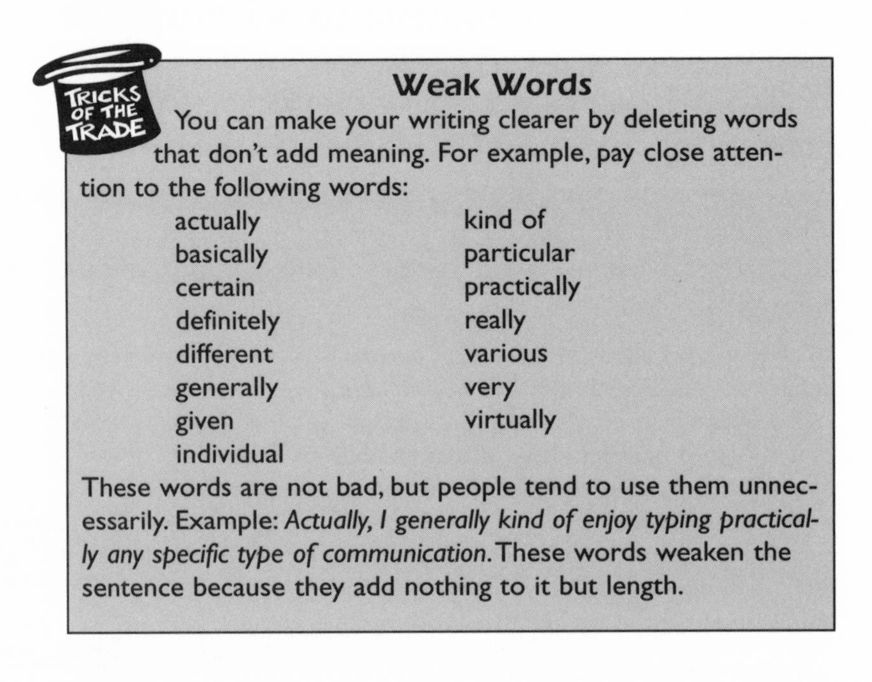

Weak Words

You can make your writing clearer by deleting words that don't add meaning. For example, pay close attention to the following words:

actually	kind of
basically	particular
certain	practically
definitely	really
different	various
generally	very
given	virtually
individual	

These words are not bad, but people tend to use them unnecessarily. Example: *Actually, I generally kind of enjoy typing practically any specific type of communication.* These words weaken the sentence because they add nothing to it but length.

Keep Sentences Short

Write in sentences of various lengths, but keep the average sentence relatively short.

If you're not persuaded that average sentence length is important, read this chart published in *communication briefings*:

Average Length	**Readability Rating**	**Readers Reached**
Up to 8 words	Very easy	90%
11 words	Fairly easy	86%
17 words	Standard	75%
21 words	Fairly difficult	40%
25 words	Difficult	24%
29 words and up	Very difficult	4.5%

How can you keep your writing lean and strong? When you might use one of the following phrases, try the shorter equivalent instead.

in the event that = if
subsequent to = after
the possibility exists for = might
prior to = before
in order to = to
in the neighborhood of = around, about
from time to time = occasionally
in reference to = about
it is necessary that = must
due to the fact that = because
in the amount of = for

Here's another example of fat writing found in a typical memo.

TO: All Employees
FROM: Brian Tinklenberg, Chief Operating Officer

This is to inform you that a new Audi, gray in color, was actually left by someone in the back parking lot. Due to the fact that the car has continued to remain there for several weeks and is still yet unclaimed, the company has been given permission by the police to auction off the car if we make a donation of the monies

> earned to some type of charitable organization. If you are willing and able to be in charge of the auction and to run the subsequent committee, please stop in at this point in time to take the time to discuss the most important and essential elements of the auction proceedings.

A possible rewrite:

> Someone left a gray Audi in the back lot several months ago. Because no one has claimed it, the police have given us permission to auction it off. We'll donate the monies raised to a charity. If you would chair the committee to run the auction, please see me.

Which memo would employees be more likely to read and understand? Both memos express the same information, but the second is stronger because it's more economical.

TOOLS

Big Impact with Little Words

Short words can often make a bigger impact. Examples:

administer = give	majority = most
aggregate = total	necessity = need
alternative = choice	opportunity = chance
anticipate = expect	peculiar = odd
articulate = explain	possibility = chance
association = group	regulation = rule
characteristic = trait	reiterate = repeat
culmination = end	requirement = need
fundamental = basic	similar = alike
indication = sign	subsequent = later
ineffective = weak	utilize = use

An easy way to write more economically, by using fewer words and keeping sentences shorter, is to reduce redundancy. How many of the following expressions do you use? How many could you reduce?

consensus of opinion/general consensus

A consensus by definition is a general solidarity of opinion.

contained herein
Contained means herein.

submitted a resignation
Use the verb—resigned.

basic fundamentals
"Fundamentals" are by definition basic.

close proximity
"Proximity" means close.

provide with information
Use the verb—inform.

cooperate together

"Cooperate" means work with others, so it's necessarily "together."

completely full
"Full" means totally or completely—unless it's not.

end result/final outcome
A "result" or an "outcome" is what you get at the end.

take under consideration
Use the verb—consider.

many in number
The word "many" can only refer to number.

future prospects
"Prospects" refers to the future; it's from the Latin, "looking forth."

sufficient enough
"Sufficient" means enough.

other alternative
"Alternative" means another choice, so "other alternative" makes sense only if there are at least three options.

new innovation
"Innovation" means new.

past experience
"Experience" usually refers to the past, so you would modify it only to refer to the present or future.

postpone until later
"Postpone" means put off until later.

true facts
"Facts" are things that are true.

mutually agree
When there's more than one party, "agree" assumes mutuality.

completely finished
"Finished" implies completely.

recurring habit
A "habit" is recurring behavior.

past memories
What else can we remember but the past?

initial preparation
"Preparation" implies initial, because it's done before something.

more preferable
"Preferable" means more desirable.

important essentials
"Essential" means important.

various different
"Various" means different. Also "different" after numbers is usually redundant: e.g., "We considered seventeen different locations" or "I called her five different times."

future plans
"Plan" implies future unless specified otherwise.

free gift
A gift is something given voluntarily, without payment in return.

continues to remain

"Remain" means to continue to be.

Straightforwardness

You can write in a more straightforward manner when you place the subject and the verb close together. Use subject-verb-object order with strong action verbs. If you write clearly, economically, and in a straightforward manner, people will find your writing more readable.

Why is it important to keep the subject near the verb? This next announcement shows what can happen when you don't.

> The executive managers of Acme Anvils, in their meeting April 5, called for the purpose of discussing problems encountered in negotiating a contract with their principal iron ore supplier, Ferrous Ingots, which has recently undergone substantial personnel changes, have decided, in consideration of the extreme importance of our iron ore supply, to arrange, as soon as possible within the limitations of their individual schedules, a meeting with the executive managers of Ferrous.

How many times did you have to read that announcement to understand what was happening? Not only is too much crammed into a single sentence, but the subject-verb-object order ("executive managers have decided to arrange a meeting") is interrupted by clauses that confuse the reader. The following revision seems more straightforward:

> The executive managers of Acme Anvils met April 5. They called the meeting to discuss problems in negotiating a contract with their principal iron ore supplier, Ferrous Ingots. The problems came out of a number of personnel changes at Ferrous.
>
> Because our iron ore supply is extremely important, the managers decided to arrange a meeting with the executive managers of Ferrous.

If we want people to read and understand what we write, we should be clear, economical, and straightforward. Now, one last comment before we move on to some other readability problems.

Order involves more than subject, verb, and object.

Sometimes managers misplace other elements of their sentences. Consider this sentence, for example: "We'll only write three major contracts this year." The position of the modifier "only" before the verb suggests that we won't do anything else but these contracts. We can make the meaning clearer if we put the modifier directly before the noun we want to modify: "We'll write only three major contracts this year."

Avoid Trite Phrases, Clichés, and Jargon

Would you like your writing to bore or confuse the people who read it? Of course not! But that's likely to happen if you use trite phrases, clichés, and jargon.

Although colloquial expressions and buzzwords can help us connect with our readers, they can also be tired or otherwise inappropriate if we simply slip them in without making conscious choices.

Here are the first two paragraphs of a cover letter I received from an individual seeking a part-time position as a writer.

> My résumé, though brief, and my little brochure pretty well tell my story. **In a nutshell**, I tired of the **corporate rat race** after over a quarter century and for seven years have been **my own man**.
>
> But as any freelancer knows, **things blow hot and cold**, and I am certainly not **booked to full capacity**. A steady thing for a couple of days a week just might be **right down my alley**.

This applicant used at least six expressions that seem worn and weak, especially for a writer. He may have wanted to come across as casual, but he seemed instead to be lazy and unimaginative. Most people skip over boring clichés, tired metaphors, and overused expressions.

In business writing you may come across phrases such as "enclosed please find," "please do not hesitate to ask," "take under advisement," "it has come to my attention," "maximum optimization," "at this point in time," and "thanking you in

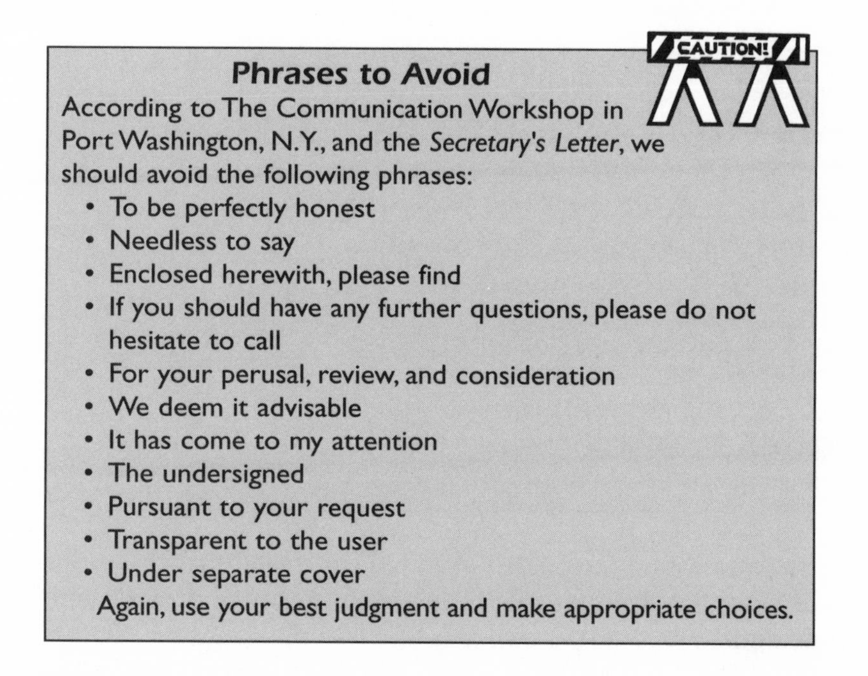

Phrases to Avoid

According to The Communication Workshop in Port Washington, N.Y., and the *Secretary's Letter*, we should avoid the following phrases:

- To be perfectly honest
- Needless to say
- Enclosed herewith, please find
- If you should have any further questions, please do not hesitate to call
- For your perusal, review, and consideration
- We deem it advisable
- It has come to my attention
- The undersigned
- Pursuant to your request
- Transparent to the user
- Under separate cover

Again, use your best judgment and make appropriate choices.

advance for your kind consideration." These are clichés that weaken your writing, and you should consciously avoid using them.

The more your writing expresses your personal presence, the more easily you can instruct, inform, or persuade the people who read your writing. In *The Careful Writer*, Theodore Bernstein offers the best advice: don't banish all clichés; use them with discrimination, not as substitutes for precise thinking.

Jargon can cause similar problems. When you use terms that are common within a particular field but not necessarily understood by people outside that field, you risk two negative reactions. The first is that your readers may not understand you—or may misunderstand you. The second is that they may feel that you don't care enough about them to connect with them, because you're using language that makes them feel like outsiders.

Watch your use of jargon. Maybe you work in publishing, so you use terms like "blueline" and "dummy" and "mechanical." Don't assume that people outside of publishing will

TRICKS OF THE TRADE

Writing for the Web

As businesses continue to expand onto the World Wide Web, it's important to know how to write well in that medium. Susan Perloff, writer for the *Philadelphia Business Journal* (Oct. 17, 1997), offers the following tips for making your Web documents more readable:

- Write short, declarative sentences.
- Involve readers by writing in the second person, addressing them directly by using "you."
- Use the active voice.
- Set off lists with bullets.
- Conceptualize the hyperlinks as you write. Remember that on the Web you never see the whole thing at once.

understand those terms. If you need to use a term that your readers may not understand, define the term or provide an equivalent.

Accentuate the Positive

The quickest way to turn off a reader is to use negative words. Research shows that negative writing decreases readability. Omit unnecessary negative information. When you must provide negative information, do so in a positive way. Sometimes giving a reason for the negative information softens the message.

Imagine receiving three letters. You open the first envelope and you read the following paragraph:

> It is essential that you comply with our request. We shall have to institute legal action against you if you do not remit the full amount of your liability by June 3.

How do you feel? Most people would react negatively. Consider the following wording of the same message:

> Please remit your payment by June 3 to avoid legal action.

The next letter also contains negative news:

> Due to an error in processing your order, it will be billed more than once to your account. A credit has been issued and we hope you have not been inconvenienced.

Mistakes happen, but it would have been better if the news had been delivered as follows:

> We apologize for the inconvenience of our processing error. We have corrected your account.

The third letter hits hard with the blunt wording:

> We regret to inform you that the merchandise you ordered is not available. Because of this, we have been forced to cancel your order.

The writer might have put the message less negatively:

> We're sorry that your merchandise is unavailable at this time.

When you can express negative messages more positively, you make your writing more readable—and your readers will react less negatively. (In Chapter 8, I discuss ways to work with negative messages.)

Positive expression is simply a matter of approaching your writing from the perspective of your readers—as I discussed in Chapter 1. Readers are naturally interested in the benefits for them and are concerned about how the negative affects them. If you focus on your readers, they'll find your writing more readable.

Make It Active

According to William Zinsser in his book *On Writing Well*, the difference between the active and passive voice is like life and death for a writer. But what is the difference?

In the active voice, the subject performs the action: "The manager wrote a memo." In the passive voice, the object is affected by the action: "A memo was written by the manager." The active voice focuses attention on the doer, while the passive voice emphasizes what's been done—allowing the writer to even omit the cause.

If you use the active voice, your writing will be more forceful, more interesting, and easier to understand. Most active verbs convey conviction and responsibility. Passive verbs hide the person responsible for the action and weaken your sentences. Compare the following constructions: "I pledge allegiance to the flag" (active) and "Allegiance to the flag is pledged by me" or "Allegiance is pledged to the flag by me" (passives). See the difference?

I'm not saying that the passive voice is always wrong. Sometimes you want to hide the responsible party: "An increase in union dues is recommended by year-end" may work better than "The union recommends an increase in dues by year-end." Sometimes the cause of an action is of little importance: "News of the merger was reported this morning in the local paper." Or sometimes the doer of an action is implicit: there's no difference between "Those terms are often misunderstood in business circles" and "People in business circles often misunderstand those terms," because we assume that the misunderstanding would be by people and not by desks or file cabinets or whatever.

On the other hand, phrases like "It was stated in the report," "It has been determined that," or "a report was sent to all divisions by the manager," and similar passive constructions take time to read and will bore readers. Researchers recommend that only 10% of your writing use the passive voice. So use active writing whenever possible and use the passive voice only when it really makes sense for what you're trying to state.

It may be difficult at first to reduce your use of the passive. Try writing your memo or report first, then editing it to remove any passive construction that doesn't work better than an active construction.

Avoid Typographical Excess

A readability problem that has been growing in recent years, with the popularity of word processing and electronic mail, is the excessive use of italics and capital letters. Research shows that the use of both italics and all capital letters (i.e., caps) makes it 20% harder for the reader to understand.

Use italics for titles of books and periodicals, foreign words, or emphasis. Don't use it just for appearances. You'll slow the reader down. *Use italics for titles of books and periodicals, foreign words, or emphasis. Don't use it just for appearances. You'll slow the reader down.* Notice the difference? And that difference increases with every line.

The use of all caps can also make your writing less readable, especially if you use acronyms. Consider the following memo from the personnel department: BRING YOUR EARS DOWN TO HUMAN RESOURCES TO CORRECT THESE PROBLEMS.

Huh? That doesn't make sense—unless you realize that EARS is an acronym for Employee Action Requests. But readers may not be able to distinguish the acronym buried among capital letters.

There's another problem with using all caps. WHEN YOU WRITE IN ALL CAPS, IT'S LIKE SHOUTING AT YOUR READERS, AND THEY DON'T LIKE IT. IT'S LIKE RAISING YOUR VOICE TO GET ATTENTION, THEN NOT LOWERING IT. YOU LOSE THE ABILITY TO CALL ATTENTION TO SOMETHING IMPORTANT, AND YOU ANNOY YOUR READERS.

Unfortunately, the use of caps has increased with the growth of e-mail. Many discussion lists advise their members to avoid using all caps. It's generally considered bad etiquette. Just keep your "Caps Lock" key off: it's as simple as that. Your readers will appreciate it—and you'll keep the power to emphasize an occasional word or phrase.

Avoid Noun and Adjective Stacks

Another way to enhance readability is to avoid stacking. Stacking means including two, three, or more adjectives in front of a noun or including several nouns in a row. This can confuse readers and slow down their reading. For example, why say "the large, ferocious, frightening wolf" when you could just use the adjective "ferocious." It implies the other two words. Most readers don't need to know that the army equipment was ruggedized, militarized, and made field accessible. They only need to know that it works in combat. Keep your writing simple. It's good for you and your readers.

Choices and Consideration

This whole discussion of readability, reduced to its essential, would probably be "choices and consideration." When you do any business writing, you make choices—medium, organization, style, tone, words, phrasings, fonts, and so on. You must understand and keep in mind the extent of your freedom to choose. Sometimes policy, company culture, or regulations might impose certain limitations on that freedom. But you should fully use whatever freedom of choice you have. Don't be lazy, negligent, careless. That's the "choices" part of the equation.

The "consideration" part is how you choose, the basis for the choices you make. You should show consideration for your readers, doing what's best to connect with them, to reach them, whether to inform or persuade or instruct. That's been the central theme of this chapter, of this focus on readability: the effect of your writing on the people who read it. That's why the title contrasts "Express" (the effect on your readers) and "Impress" (the benefits for your ego).

Now that you know the many ways to make your writing more readable, consult the following checklist to ensure that you write to express, not impress.

Manager's Checklist for Chapter 3

- ❑ Write clearly. Make sure that your readers understand you.
- ❑ Use economy with your words.
- ❑ Be straightforward, placing subjects near verbs, minimizing distracting clauses and phrases.

- ❑ Avoid trite phrases, clichés, and jargon.
- ❑ Accentuate the positive when delivering negative messages.
- ❑ Emphasize the active voice over the passive.
- ❑ Avoid excessive use of italics, and eliminate writing in all caps.
- ❑ Eliminate stacking of adjectives and nouns.
- ❑ Keep in mind "choices and consideration." Use your freedom fully and consciously and think from the perspective of your readers.

Structure Your Writing to Reach Your Reader

When writing e-mail, memos, letters, and reports, make sure you ask for what you need and explain why you need it. In this chapter, I discuss several easy formats for internal and external communication—letters and memos to make requests, to inform, to persuade, and to deliver good news and bad news. In most cases, the principles will serve you well and provide structure for your reader.

To write letters and memos that will capture and hold your reader's attention, you must mesh strategy and structure. We look first at a few techniques, discuss some principles of business communication, and then consider five step-by-step approaches to writing more effective and efficient letters and memos.

Techniques to Engage the Reader

Chapter 1 discussed how you can know your readers and build rapport. Chapter 2 emphasized focusing on your purpose and on pursuing that purpose through the four phases of planning, writing, revising, and editing. Chapter 3 covered various ways to put more power into your words. Now, as we begin to discuss strategies and structures, we face a crucial question: How do you get somebody to read your writing?

The first point is to remember that business truism: you never get a second chance to make a first impression. Appearances matter in business writing, so make sure your memos and letters are neat and well laid out, with sufficient margins all around and paragraphs spaced for easier reading. (I'll provide samples to show you what I mean.) With the quality of printers in business use now, you should never have any problems with the appearance of your words as long as you choose a legible font and an appropriate point size.

But you can also do several other things to get people to read what you write.

Entrance and Exit Ramps

Make it as easy for your readers to get into your writing as to get out of it. Indent the first line of each paragraph to draw your readers into the text, like the on-ramp allows easy access to a highway. (Note: Books often do not indent the first paragraph under a heading, as is the convention in this book. All other paragraphs are indented.)

Make your first few paragraphs short to give your readers an easy, early exit from your writing. They're more likely to enter into it if they see that they don't have to stay in it for a long time before they have a chance to exit.

Subject Lines and Postscripts

Use subject lines and postscripts as additional inducements for your readers. A good subject line reveals the topic of the e-mail, report, memo, or letter and can motivate a recipient to at least skim the first paragraph. In written communication such as letters, memos, and e-mail, a postscript can grab attention. In fact, many people read the postscript first. It can be a powerful addition, especially for persuasive documents. Use it to hit your most important point.

Now, let's apply these techniques in letters and memos.

Traditional Letter Form

Let's start with the basic letter format. Because most managers today use computers, many also format their own correspondence. Letters have two basic formats, the block and the modified block. The block format gives letters a more formal look; the modified block format is more like the format for a personal letter. Consider the examples in Figures 4-1 and 4-2.

Lord Consulting Services
699 Knox Road
Ardmore, PA 19077
(610) 555-0124 — **Heading**

January 23, 1998 — **Date**

Mr. James Penrod
Vice President, Sales — **Inside Address**
ABC Computing
394 Vesper Road
Knoxville, TN 37966

Subject: Megasoft Word Pro 9.1 — **Subject line**

Dear Mr. Penrod: — **Salutation**

Body

I would like to order the most current edition of Megasoft Word Pro, which I believe would be the 9.1 version. Would you please send me any appropriate documentation as well? I work at home as a consultant and prepare brochures and newsletters for my clients.

Because I need your product immediately, please send it in the quickest way, perhaps by overnight mail. I have always used your software products and appreciate your service mentality. Thank you for handling this request quickly.

Sincerely yours,

Judy Lord — **Closing and signature**

Judy Lord
President

ALD/jl.245 — **Supplement line**
Attachments (2) — **Attachments**

P.S. I enjoyed your holiday newsletter! — **Postscript**

Figure 4-1. Modified block format letter

Lord Consulting Services
699 Knox Road
Ardmore, PA 19077
(610) 555-0124 — **Heading**

January 23, 1998 — **Date**

Mr. James Penrod
Vice President, Sales
ABC Computing
394 Vesper Road
Knoxville, TN 37966 — **Inside Address**

Subject: Megasoft Word Pro 9.1 — **Subject line**

Dear Mr. Penrod: — **Salutation**

Body

I would like to order the most current edition of Megasoft Word Pro, which I believe would be the 9.1 version. Would you please send me any appropriate documentation as well? I work at home as a consultant and prepare brochures and newsletters for my clients.

Because I need your product immediately, please send it in the quickest way, perhaps by overnight mail. I have always used your software products and appreciate your service mentality. Thank you for handling this request quickly.

Sincerely yours,

Judy Lord — **Closing and signature**

Judy Lord
President

ALD/jl.245 — **Supplement line**
Attachments (2) — **Attachments**

P.S. I enjoyed your holiday newsletter! — **Postscript**

Figure 4-2. Block format letter

Use a heading, including your phone number, even if you're writing from home, so that the reader can easily identify you or reach you, if necessary. Always include the date. The inside address fulfills two purposes: first, you can use the person's title, which most people like to see; and second, it routes the letter to the appropriate individual, even if someone else opens the mail.

The subject line specifies the purpose of your letter. The salutation (followed by a colon, not a comma) identifies the recipient. Always try to put a name rather than just a position title: you're more likely to get your recipient to read your letter if you don't use the business equivalent of "Occupant." If you cannot obtain a name, use Dear Student, Dear Customer, Dear Homeowner, or some salutation that identifies the recipient in some way. Use Dear Sir/Madam or To Whom It May Concern only as a last resort. Never use Gentlemen unless you are sure that no woman may read the letter.

Keep your first paragraph short to intrigue the reader. Indent each paragraph for reading ease. Use one of the many good closings to end your letter: Cordially, Sincerely, Sincerely yours, Regards, Respectfully submitted, and so on.

The supplement line is for your benefit; it indicates who formatted the final letter and where you can find this letter on a computer disk. If you plan to enclose a check or other document, indicate how many attachments you're including, to alert the recipient to the enclosed items.

The postscript gives you a final chance to catch the reader's attention. It's a good place to remind the reader of your main point or to call attention to a deadline.

Memo or E-Mail Form

Most managers use memos for internal issues and letters for external communication. In many organizations, e-mail has replaced memos because of its speed and ease of use.

Minding Your Electronic Manners

According to Barbara Pachter, a communications trainer, sloppy e-mail shows bad business manners. She suggests these guidelines:

- Don't contribute to e-mail overload.
- Keep your message short (one screen or 25 lines).
- Use short paragraphs.
- Use a subject line.
- Don't use all capital letters.
- Limit each message to one subject area or purpose.
- Proofread each message.
- Remember that e-mail is not private.

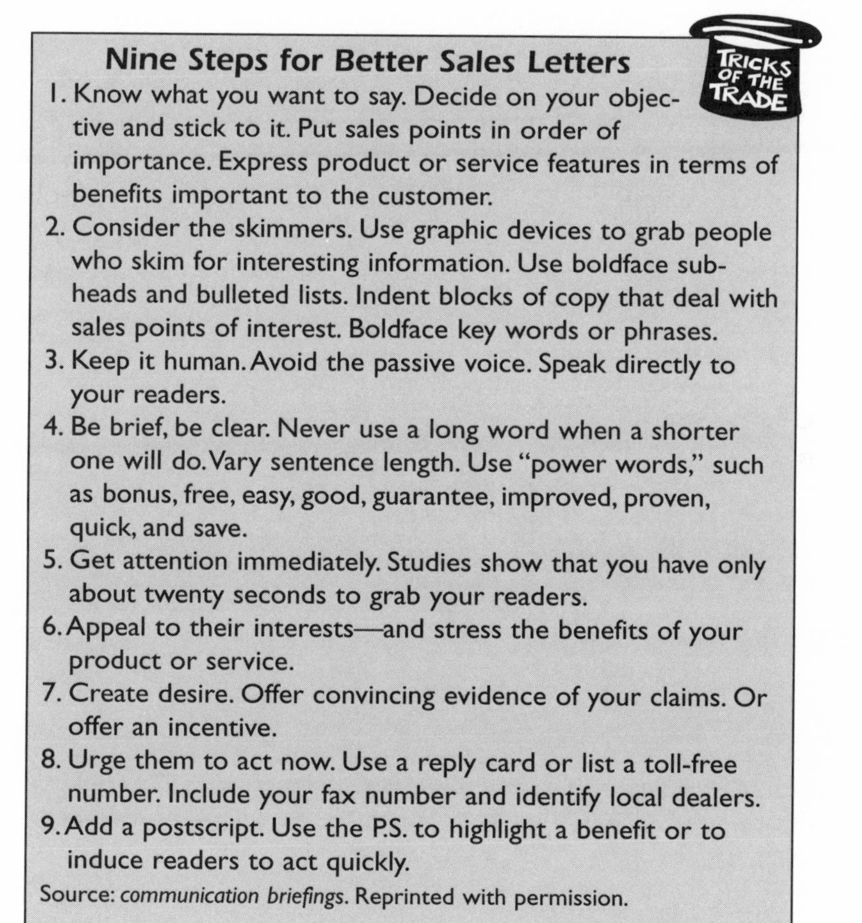

Nine Steps for Better Sales Letters

1. Know what you want to say. Decide on your objective and stick to it. Put sales points in order of importance. Express product or service features in terms of benefits important to the customer.
2. Consider the skimmers. Use graphic devices to grab people who skim for interesting information. Use boldface subheads and bulleted lists. Indent blocks of copy that deal with sales points of interest. Boldface key words or phrases.
3. Keep it human. Avoid the passive voice. Speak directly to your readers.
4. Be brief, be clear. Never use a long word when a shorter one will do. Vary sentence length. Use "power words," such as bonus, free, easy, good, guarantee, improved, proven, quick, and save.
5. Get attention immediately. Studies show that you have only about twenty seconds to grab your readers.
6. Appeal to their interests—and stress the benefits of your product or service.
7. Create desire. Offer convincing evidence of your claims. Or offer an incentive.
8. Urge them to act now. Use a reply card or list a toll-free number. Include your fax number and identify local dealers.
9. Add a postscript. Use the P.S. to highlight a benefit or to induce readers to act quickly.

Source: *communication briefings*. Reprinted with permission.

Unfortunately, the ease with which we can write e-mail messages has led to problems. Many e-mail users tend to write as casually as they talk, forgetting that e-mail messages are a written form of communication and they last far beyond a casual conversation. A good manager puts thought and care into writing e-mail messages.

Let's take a look at the basic memo or e-mail format (Figure 4-3, page 54).

Use a standard memo format with the following headings: TO, FROM, DATE, and RE or SUBJECT. Use a colon after each capitalized heading and double-space them. Remember to initial the memo, indicating that you've read it and that it

TRICKS OF THE TRADE

E-Mail Etiquette (aka Netiquette)

Marjorie Brody, president of Brody Communications Ltd., offers the following tips on e-mail etiquette:

- Watch your words.
- Be concise, to eliminate follow-up phone calls.
- Don't make antagonizing or critical comments ("flames").
- Few people like "spam"—unsolicited e-mails.
- Nothing is private. Anybody may read your messages.
- Keep attachments to a minimum: they can overload systems and tax patience.
- Consider copying others in the office, using the FYI or For Your Information designation.
- Never assume anything. Many users aren't familiar with the lingo, emoticons, and e-mail acronyms that you may be using.

TO: All Employees
FROM: Human Resources
DATE: March 1, 1998
SUBJECT: Holiday policy

Because our employee committee suggested a more flexible holiday policy, we have adopted several options. You may now choose either Columbus Day or President's Day as a paid holiday.

Previously, you could select one of these days as a floating holiday.

Now you may select one as a paid holiday.
If you have questions about the policy changes, call Zack at 4356.

c: Board of Directors
P.S. Enjoy your extra holiday!!

Figure 4-3. Basic memo or e-mail format

contains what you want it to say. Check content, grammar, and spelling.

Traditionally, the term "cc" was used at the end of a memo to identify those who would receive carbon copies. Because the carbon copy is extinct, most writers now use just "c" for "copy."

Consider using a postscript in your e-mail or memo. It catches readers' attention, causing them to look at the subject line. You then have another way to lure the reader into your memo.

When to Copy Someone

MISTAKE PROOFING

Copy an employee's boss, particularly if your memo cites the employee's accomplishments. When employees receive that kind of recognition, they tend to work harder. Also, managers often copy employees on an FYI, or For Your Information, basis.

Purposes for Memos

Memos are an ideal medium to signal a shift in position. You can make important points quickly in a single page. It's generally best to put the change in your opening paragraph—unless the nature of the change suggests that it might be wiser to buffer the announcement with a paragraph or two to provide a context and justification.

Memos also can provide valuable information. Be concise; don't exceed one page. Identify your subject in a way to grab attention. It's also effective to sum up the information in a brief postscript.

Use memos to describe action taken or planned and to serve notice of deadlines. Especially in the case of actions, it's a good idea to send an FYI copy to people who might be indirectly affected.

Memos provide an easy way to summarize meetings, particularly those without minutes, and are a good way to keep employees current about what's happening behind closed doors. After a morning meeting, write up a memo about what happened and have it waiting for participants and others when they return from lunch. If the meeting takes place in the afternoon or evening, you can get a memo out the first thing in the morning. Informing everyone immediately is a good way to reduce inaccurate word-of-mouth accounts.

Content and Form for Memos

Make your memos a single page. Many executives refuse to read any memo that exceeds one page. If that's your situation (or even if it's not), this suggests that you should provide only the essential information. You can always invite those interested in details to contact you. A tip: write up your memo without regard for length, then use that memo as the basis for a brief memo and keep it around as a follow-up for those who want more information. Follow the B-E-T method:

- **Bottom line.** Start by stating your conclusion or action. Get to the point immediately.
- **Evidence.** Provide the reasons for your conclusion or action.
- **Tasks.** Identify what should be done, who should do it, when, how, and why.

Write in the second person to establish the most direct connection with your readers. Use "you" as much as possible to involve them in the action.

Tips for Writing at Chip Speed

Our high-tech, high-speed business world demands to-the-point writing that gets the message across effectively and efficiently. According to Jack Gillespie, editor of *communication briefings*, even when we send e-mails or faxes, we still need good writing techniques. He advises the following:

- Use simple sentences.
- Keep your sentences short, an average of 17 words or fewer.
- Address your readers with "you."
- Use active verbs.
- Avoid using nouns for verbs. (Instead of "give a description of," why not just "describe"? What advantage is there in using "take into consideration" rather than "consider"?)
- Strive to make your message clear.
- Use short paragraphs with subheads.

In other words, the principles that help you write more effective and efficient letters and memos make even more sense in shorter, faster business media like e-mail and faxes.

Five Formats for Letters and Memos

Now that we've reviewed the basic letter and e-mail format and how to write "at chip speed," let's look at five possible structures for letters and memos.

Direct Request

When writing a direct request, the two most important pieces of information you should include are why you need the requested item or service and how you will use it. Your reader will usually welcome the direct request. The reader's attitude is positive; he or she wants to hear from you. Try this format:

Paragraph 1. Request the information or services you need.

Paragraph 2. Show why you need the information and how you will use it.

Paragraph 3. Specify what you want the reader to do.

Paragraph 4. List the benefits for the reader and use a good-will ending.

For example, if you write to a software company asking to purchase a program, the most important information you can give the company is why you need the software and how you'll use it (Figure 4-4, page 58).

This direct request letter consists of four parts. Susan Harris requests the magazine samples and advertising information. She explains why she's requesting the samples and information and how Summa Publishing will use the samples. (The director of marketing at Pan Periodicals will then know what other information might be appropriate to send.) She indicates when she'll need the samples and suggests how the director of marketing should send them. She closes with an indication of the business potential of this contact, the possibility of benefits for Pan Periodicals.

That's it: the essentials of a direct request in four steps.

Informative

We often write letters, memos, or e-mails to provide information about upcoming meetings, policies, or projects. Usually

Summa Publications
2718 North Port Drive
Kalamazoo, MI 49001
(616) 555-0154

July 10, 1998

Ms. Joanne Meyers
Director of Marketing
Pan Periodicals
123 Seventh Avenue
Madison, WI 53712

Subject: magazine samples

Dear Ms. Meyers:

I would like to request samples of your magazines. Please also send me your advertising rate sheets as well as any other information you might believe appropriate.

We at Summa Publications are expanding our marketing efforts for our specialty newsletters from direct mail to space advertising. We believe that some of your magazines may be good vehicles for our promotional efforts. I'm enclosing a brochure that lists and describes our newsletters.

Because we will be meeting to discuss our space advertising campaign July 29, we'd like to receive your samples by July 22. Please send them by first-class mail.

We intend to start devoting a large percentage of our promotional budget to space advertising in 1999. We believe that Pan Periodicals could be a good business partner with Summa Publications in the years to come. Thank you for handling our request promptly.

Sincerely yours,

Susan Harris

Susan Harris
Assistant Director, Product Promotions
Summa Publications

PAV/req.001
Attachment

Figure 4-4. Direct request letter

readers have a neutral attitude toward informative letters, so the most important aspect of this letter is to capture their attention.

Paragraph 1. Capture attention.
Paragraph 2. Provide the necessary information.
Paragraph 3. Present any negative factors; show reasons for these factors.
Paragraph 4. List the benefits for the reader.
Paragraph 5. Close with a goodwill ending.

Start by capturing your reader's attention. Remember the basic question, as discussed in Chapter 2: "So what?" Why should the person receiving your communication care about reading it?

Then give the necessary information. If there are negative factors, embed them among more positive paragraphs. For example, if an employer designates a room as a smoking lounge, a negative factor could be limited times when smokers could use the room. Never start or end with negatives!

Always try to list the benefits for the reader, to leave the reader feeling better about your announcement. Then close with a goodwill ending. Let's look at an example of an informative memo (Figure 4-5, page 60).

Clarence Littlejohn attracts attention from the start by announcing a benefit for the employees. He mentions a negative factor—a week of inconvenience. (There's no need in this case to give reasons for this negative factor.) He reminds the employees of the benefit for them, then concludes with a positive, enthusiastic ending.

Persuasive

Sometimes we must not only inform but also persuade our readers. That requires a little more skill. Try the following format to persuade a reader who may initially resist your request.
Paragraph 1. Catch the reader's interest; establish mutual goals or common ground.
Paragraph 2. Define the problem that will be solved if the reader approves your request.
Paragraph 3. Explain the solution and show how the advantages of the solution outweigh any negatives.

TO: Employees of Parkdale Pens and Pencils
FROM: Clarence Littlejohn, Facilities Manager
RE: Company Parking Lot
DATE: August 10, 1998

I'd like to announce that the executive board has approved plans to resurface the company parking lot.

We'll begin work on August 17. We expect the project to take about a week.

During that time, no one will be able to use the parking lot. Let's put up with what I hope is a relatively minor inconvenience, and after the repairs we can all look forward to a parking lot without potholes and puddles. We'll have more room and the spaces will be more clearly marked.

The work should be complete by August 24.

I hope you share my enthusiasm about this improvement of our facilities.

Figure 4-5. Informative memo

Paragraph 4. List all the benefits for the reader.
Paragraph 5. Specify what you want the reader to do.

Even if the reader might disagree with you, try to establish mutual goals, agree on some point, or establish common ground. Involve the reader by presenting a problem with multiple solutions. Most people like choices; the multiple solutions allow the reader an opportunity to choose.

If you must give any negative information, make sure the advantages of the solution outweigh the negative factors. List any benefits for the reader that would result from his or her solution to the problem. And most important, state the specific action you want the reader to take. Many times a reader feels persuaded by the message, but doesn't know what to do next. Figures 4-6 and 4-7 are examples of persuasive messages

First, Lindsay McHugh captures the attention of the employees by using a startling and scary fact. Then she

TO: Employees at Ross Technical Tools
FROM: Lindsay McHugh, President
RE: Organ Donor Awareness Celebration
DATE: September 17, 1998

Every eight minutes someone dies waiting for an organ. We need your help to participate in a special event to help these dying mothers and fathers, grandparents, children, and even babies receive a second chance at life.

Some of you may know someone who needs an organ or who has received an organ from a stranger or loved one. Let's work together to help people in need in our community.

Join us on October 28 by stopping by the break room between 10:00 and 2:00 to enjoy some delicious snacks and to pick up the informational material on display there. We'll also have a special guest there to answer any questions—Kirk Seymour, a research assistant in R&D here at Ross, who just last year received a liver from his sister.

It's easy to donate organs. You simply sign the donor card. Then, upon your death, doctors will be able to use your organs—if your family members consent. You can also donate an organ while you're alive. If you closely match someone in need, you can donate an organ to save a brother or a sister, maybe a son or a daughter.

Please join Kirk and me in the break room between 10:00 and 2:00 October 28 for chips, veggies, cheeses, and cookies—and a chance to save somebody's life.

Thank you!

P.S. Just drop in for a few minutes. Your family may need to count on you.

Figure 4-6 Persuasive memo

defines the problem: people in the community are dying because they need an organ transplant. She explains how this event can provide a partial solution to a horrible problem by making employees aware of how they can donate their organs.

Because many people are reluctant to sign donor cards or even to think about dying, the president next invites the employees to enjoy some snacks, following that benefit with a mention of the informational materials that they can take. She personalizes the need for transplants by featuring a Ross employee who received an organ. In that context, the president briefly explains the use of donor cards, referring to death only in passing and then ending the paragraph with a mention of loved ones. In other words, she subtly suggests that the advantage of saving a life outweighs the disadvantages of donating an organ.

She closes with a sentence that both sums up the benefits for the employees and tells them what they should do. She then thanks them in advance and adds a postscript that makes a final appeal, casual yet urgent.

Bell Atlantic Business Systems
1100 Carriage Drive
Tonawanda, PA 19087
716 / 555-0101

November 9, 1998

Matthew Hipple
Maryland Unemployment Insurance
1100 N. Eutaw Road
Baltimore, MD 21201

Subject: Third quarter return, Account number 0090692510

Dear Mr. Hipple:

[get attention] We have enclosed our third quarter 1998 unemployment return. As you can see by the date stamped on the envelope, we mailed our return in a timely manner to an address given to me by someone in your office. **[define problem]** I now understand that the address has not been valid for several years.

Because we have always filed our returns, including this one, in a timely manner and mailed them in good faith, **[explain solution]** we would like to request that any late penalties or interest be abated. **[give benefit]** Our accounts payable department now has your new address, and in the future we will send all payments to your new offices.

[**state action**] Please let me know our current status with you. We hope that you will decide not to assess any late charges or fees because of this mailing error. Thank you for your understanding.

Sincerely,

Marco Umberto

Marco Umberto,
Payroll Tax Accountant

Figure 4-7 Persuasive letter

Good News

And now for the letter or memo that's easy to write, because everybody likes to get good news. This communication has four basic steps:

Paragraph 1. Deliver the good news.
Paragraph 2. Provide any details.
Paragraph 3. Discuss any negative elements.
Paragraph 4. List the benefits for the reader and close with a goodwill ending.

Take advantage of opportunities to applaud a colleague or an employee by sending good news memos on any deserving occasion. Research shows that employees perform better when they feel appreciated and recognized.

Start your memos by announcing the good news, then follow up with any details. Touch on any negative elements after detailing the good news. Then return to the positive by listing the benefits for the reader. Close with a goodwill ending. Figure 4-8 (page 64) shows an example of a good news letter.

Good news letters and memos can be a lot of fun to write—and they can improve motivation and increase business. Unfortunately, many managers fail to take advantage of the opportunity to communicate good news.

Negative Message

The negative message is perhaps the most difficult letter or memo to write. You and your reader will naturally dread this kind of letter or memo, so you've got to be particularly careful

Avery Bedding Company
303 Queen's Way
Boothbay, ME 77204
207 555-0027

May 14, 1997

Ariel Bradford, Purchasing Manager
Beds R Us
101 Stockton Court
East Aurora, NY 14052

Dear Mr. Bradford:

We're pleased to announce that one of our sales representatives, Samantha Malone, has been chosen Seller of the Year by the National Bedroom and Bath Association.

You've worked with Sammy for several years now and you know how she cares that everybody gets a good night's sleep. The NBBA is now recognizing what we at Avery Bedding Company have long known, that Sammy is the best in the business.

There's some bad news: Sammy has to fly to Morgantown, WV, next week to accept the award, so she won't be able to make all her calls this month. We hope that won't inconvenience you.

You know Sammy: she'll be out there helping you serve your customers as soon as she returns. And I'm sure you'll agree with me that this award will only inspire her to work even harder to meet your needs for the best bedding in the world.

Please call me if you have any questions before Sammy makes her next visit.

Sincerely,

R.J. Avery

R.J. Avery
Vice President of Public Relations

Figure 4-8. Good news letter

in writing it. I suggest the following three steps:

Paragraph 1. Establish goodwill.

Paragraph 2. Present the negative message. Give reasons for the message.

Paragraph 3. Explain the positive aspects and reestablish goodwill.

Start by establishing common ground or goodwill to prepare the reader for the bad news. Then deliver the negative message. Give reasons, if possible: this may be the most important part of your letter or memo. To close, explain any positive aspects and reestablish the goodwill of your reader.

Most managers have occasion to write negative message letters when they're hiring for a position. It's easy and enjoyable to notify the candidate who receives the position, but you've also got to inform the other candidates. Figure 4-9 presents an example of this type of negative message letter.

Jonathan Henry begins by emphasizing the positive experience of the interview and praising the applicant for her enthusiasm and knowledge of the industry. He then delivers the negative message, but tactfully, in positive terms: the deci-

Drug Depot
3449 Washington Boulevard
Portage, MI 49081
616 / 555-0187

September 2, 1998

Grace O'Dowd
1704 Center Street
Portage, MI 49082

Dear Grace:

Thank you for interviewing with us last Wednesday. We enjoyed your enthusiasm for your field and your knowledge of the pharmaceutical industry.

Unfortunately, we have decided to offer the sales position to a candidate with more experience in the field. However, we may have openings in the next four months for additional account executives.

We will keep your résumé on file and contact you when these additional positions open up. In the meantime, we wish you success with your career search. Thank you again for thinking of Drug Depot.

Sincerely,

Jonathan Henry

Jonathan Henry, Sales Manager

Figure 4-9. Negative message letter

sion is not a rejection of Grace O'Dowd but the selection of another applicant.

He gives a reason for the decision: the selected applicant is more experienced. In a letter to a job applicant, it's generally wisest and kindest to emphasize positive reasons for selecting another applicant, rather than finding fault with the applicant who is not selected.

Henry follows up his negative message with a positive: positions may open up soon and he invites her to apply for them. He promises to keep her résumé on file and contact her. This promise is too often just standard procedure, so if you make such a promise, be sure to follow through on it and contact the applicant when a future opening arises. Henry closes with best wishes and a final expression of appreciation.

It's tough to notify job applicants who don't come out on top, but a good manager can ease the pain of the disappointment with a good letter.

Let's look at another example (Figure 4-10) of establishing goodwill, presenting the negative message, and reestablishing goodwill.

Simplify with Structures

This chapter has focused on using formats for letters and memos to make requests, to inform, to persuade, and to deliver good news and bad news. Structures such as I've outlined here can take you step-by-step through the essentials of each communication situation. These structures can simplify life by making it easier for you to write letters and memos and easier for the recipients of your messages to read them.

Manager's Checklist for Chapter 4

- ❑ Get your readers into your writing by being careful and thoughtful with appearances, indenting your paragraphs, making your opening paragraphs short, and using subject lines and postscripts.

Nadia Metropolitan Interiors
133 Paree Drive
McKeesport, PA 15563
412 555-0101

July 1, 1998

David Dougherty
Decors by David, International
7 Oakmont Place
Pittsburgh, PA 17085

Dear Mr. Dougherty:

[establish goodwill] Thank you for your recent shipment of the furniture we ordered. We've always enjoyed dealing with your company and we're proud to sell your products.

[deliver negative message] However, as I mentioned to you on the telephone yesterday, we're disappointed by the condition of the six white mushroom chairs in this shipment. I've enclosed two photographs of the chairs that show how the paint is flaking off in many areas. It appears that the paint was applied over a finish and so is not adhering properly.

Please replace these chairs. If you cannot guarantee the white finish, I will accept the natural finish as a substitute.

[reestablish goodwill] I know how much you care about customer concerns and I look forward to a resolution of this problem.

Sincerely yours,

Nadia Carlson

Nadia Carlson
President

Figure 4-10. Negative message letter

- ❑ Structure your writing to reach your reader.
- ❑ Follow traditional formats for letters and memos so your readers know where to find what they need.
- ❑ Mind your manners when using e-mail. Treat them at least as carefully as you treat memos and letters on paper. Follow the guidelines of "netiquette."
- ❑ Follow the five step-by-step approaches to writing letters and memos to make requests, inform, persuade, and deliver good news and bad news.

The Power of Visuals, White Space, and Headings

Many writers forget how readers naturally react to looking at an entire page of text with no breaks. To appeal to your readers, you should use visuals, white space, and headings. In this chapter, we examine ways to get the most out of these elements to grab and hold the attention of your readers.

Visuals: Into the Head Through the Eyes

Visuals can be a great way to attract attention to the text, to emphasize main points, and to present important information. Most of us are attracted to graphics, whether photographs or designs, whether in color or in black and white. We tend to look at the graphics before we read the text.

Think, for example, about *USA Today*. That newspaper with its graphics changed the traditional look of newspapers. However you may feel about the info bites presented, you cannot ignore the fact that most people are immediately drawn to them.

The pictogram shown in Figure 5-1 is an adaptation of a

USA Today graphic illustrating the most used World Wide Web sites. Notice how quickly you can get the message without much text.

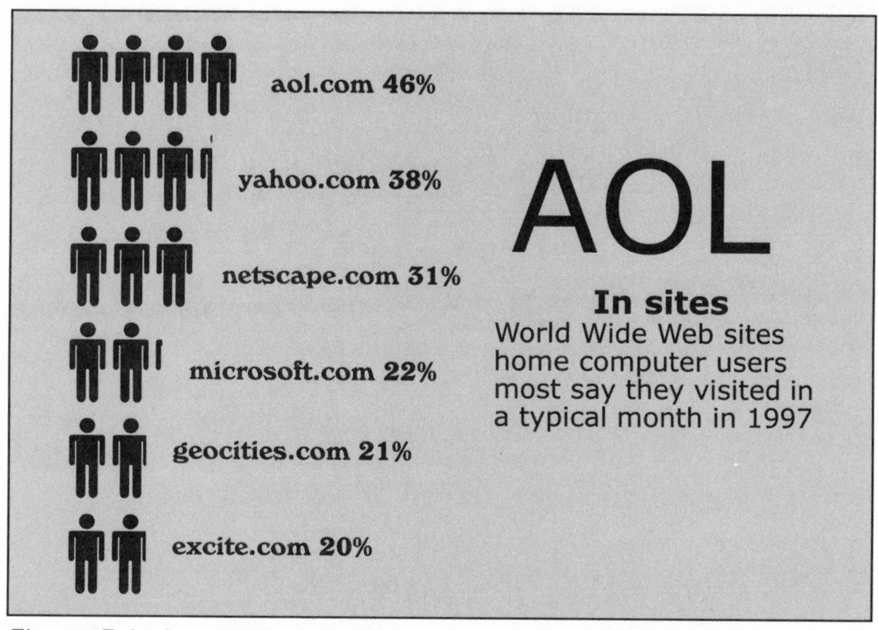

Figure 5-1. An adaptation of a pictogram from **USA Today**

We've all heard that a picture is worth a thousand words. That's not necessarily so. It's important to use visuals appropriately. Good visuals are worth at least hundreds of words. Bad visuals can distract and detract from your text and can undermine the impact of your writing as much as if you had used no visuals at all.

When to Use Visuals

Use visuals to accomplish any or all of the following purposes:

- To support text
- To convey information
- To direct action

Visuals can support text by clarifying a confusing or complicated point or by reinforcing a concept. For example, if you're quoting many percentages, the figures could confuse

> **For Example**
>
> **Pictograms**
>
> *USA Today* uses pictograms or visuals to present information in ways that are interesting graphically. The information may not be particularly important in itself. (Do we really care about comparisons of snowfall in the last five winters as depicted by various sizes of snow shovels?) But the graphics pull us into the articles by breaking up the text and making it more accessible. That's important to remember when you're preparing a report.

your reader, so also use a pie chart to depict those figures graphically and emphasize proportions.

Visuals can convey information more powerfully than text can. For example, tables can present a lot of related information with greater impact than if you were to use words to present the information and explain what it all means.

You can direct action with a visual that refers the reader to a diagram or flowchart that describes a process or explains how to follow a set of instructions.

How to Use Visuals

Michael Keene, author of *Effective Professional Writing*, recommends the following three guidelines for using visuals to attract attention:

- Accessibility
- Appropriateness
- Accuracy

First, make sure you locate your visual somewhere near the text to which it relates, facilitating reader access. Identify the visual, provide a short explanation, and frame it with white space. Readers find it annoying when they have to page through a report to find a chart or a graph. Make it easy for readers to find and read your chart or graph and at the same time refer to the text.

Second, you should use a visual that is appropriate to your purpose. The following list presents the basic advantages and disadvantages of the most common visuals:

- **Tables.** Good for conveying a lot of numbers but not for emphasizing key points.
- **Pie graphs.** Good for showing percentages of a whole but not for presenting complex data or patterns over time.
- **Bar graphs.** Good for comparing several items, especially over time, but not for showing more than ten items or showing divisions of a whole.
- **Line graphs.** Good for showing a pattern over time but not necessarily for showing factors that affect that pattern, so these graphs can become confusing.
- **Pictograms.** Good for attracting attention creatively but not for making highly accurate comparisons.
- **Flowcharts or diagrams.** Good for showing a complex process but sometimes confusing when they depict complex or extensive processes.
- **Photographs.** Good for setting a mood or conveying a sense of realism but annoying if the focus is off or the composition is wrong.

Choose your visuals carefully in terms of your information and your purpose. Consider the comparisons of pie chart and bar graph (Figures 5-2 and 5-3, pages 72 and 73), stacked bar graph and stacked area graph (Figure 5-4, page 74), and bar graph and line graph (Figure 5-5, page 75).

The Right Visual

MISTAKE PROOFING

To reiterate, use the visual most appropriate to your information:

- Tables—to convey a quantity of statistical data (numbers).
- Pie graphs—to show percentages of a whole, not complex data.
- Bar graphs—to compare ten or fewer items.
- Line graphs—to show a pattern over time.
- Pictograms—to attract the reader's attention creatively, not for highly accurate comparisons.
- Flowcharts or diagrams—to show a complex process.
- Photographs—to set a mood or to show an actual situation or person.

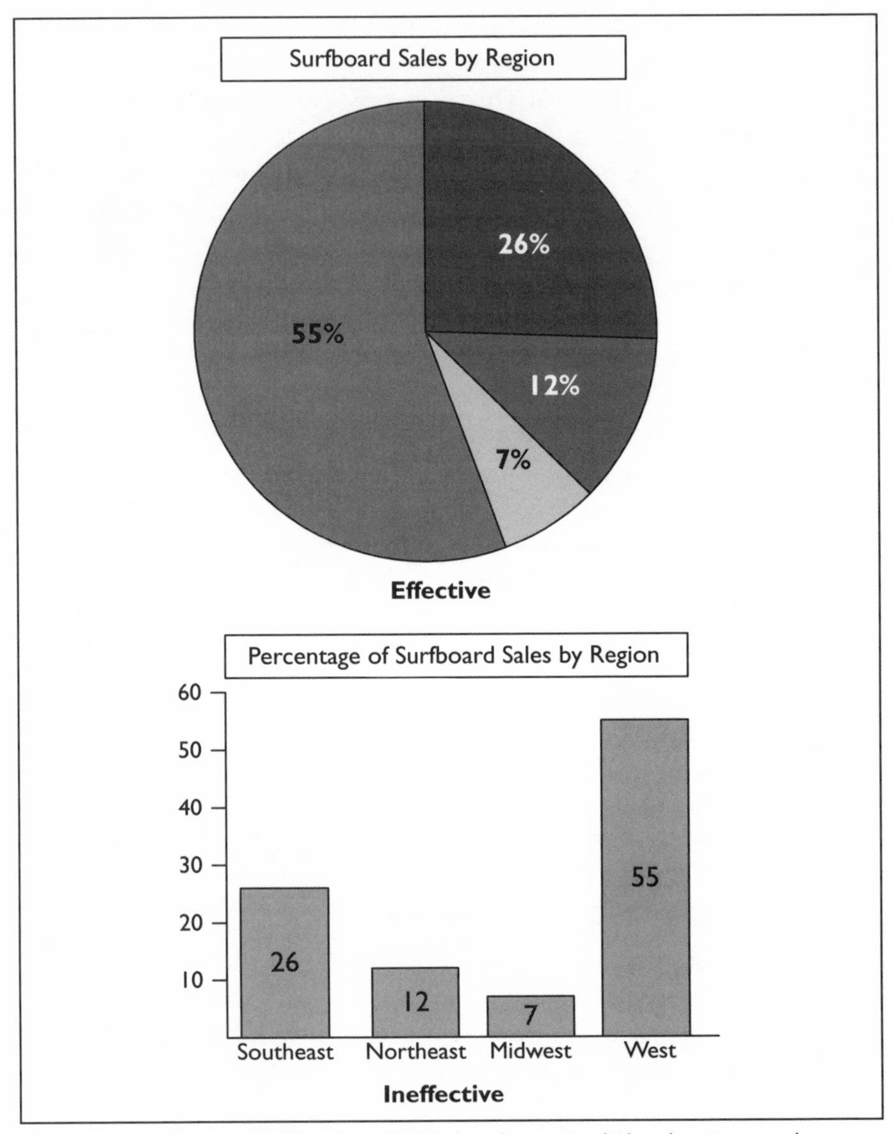

Figure 5-2. The top diagram is a pie chart and the bottom a bar chart. A pie chart is an effective way of showing percentages. The whole pie equals 100%, and each slice represents the percent allocated to that slice. The same data in a bar chart does not effectively show that selected items are parts of a whole equaling 100%.

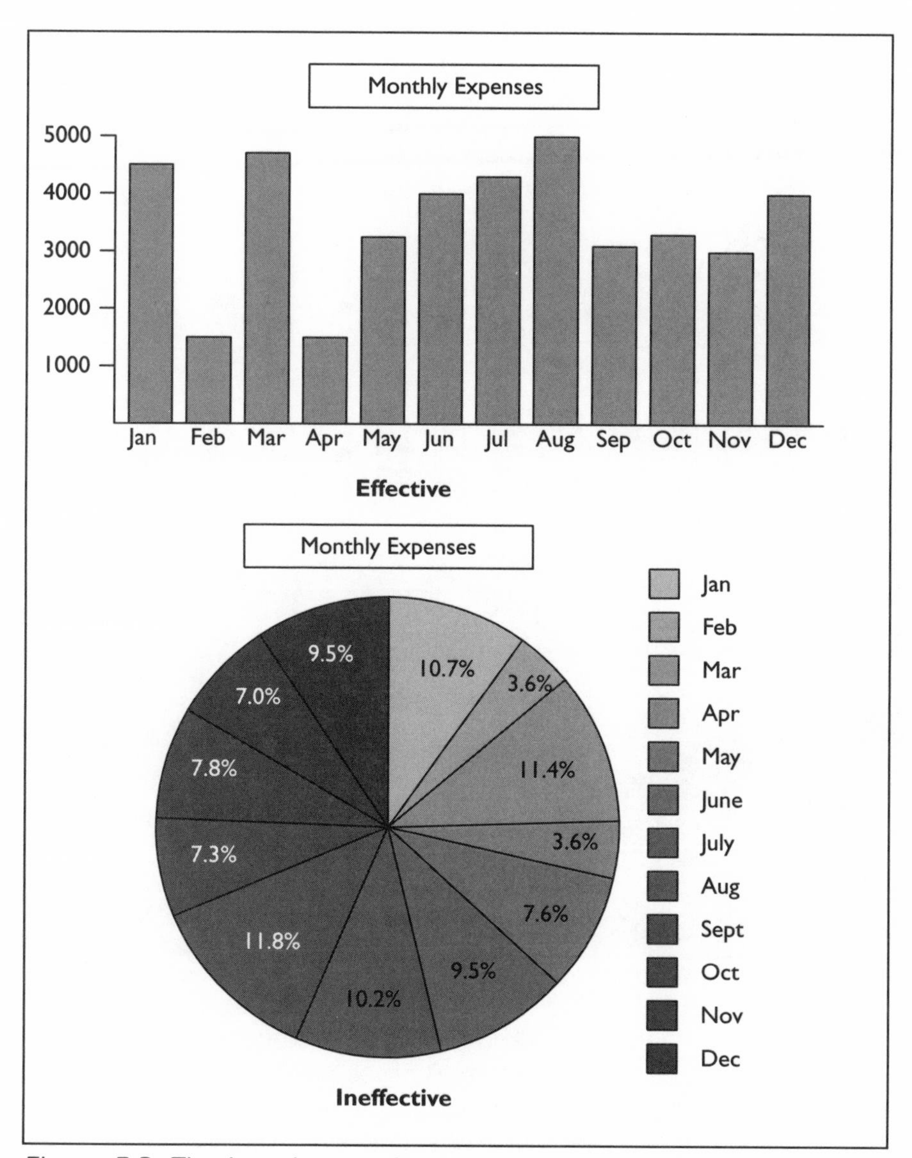

Figure 5-3. The bar chart at the top is an appropriate way to compare data over time and determine trends. The bottom figure shows similar data in a pie graph form, with the months broken out as percentages of the entire year's expenses. Notice that in the pie chart the trend over time is less obvious.

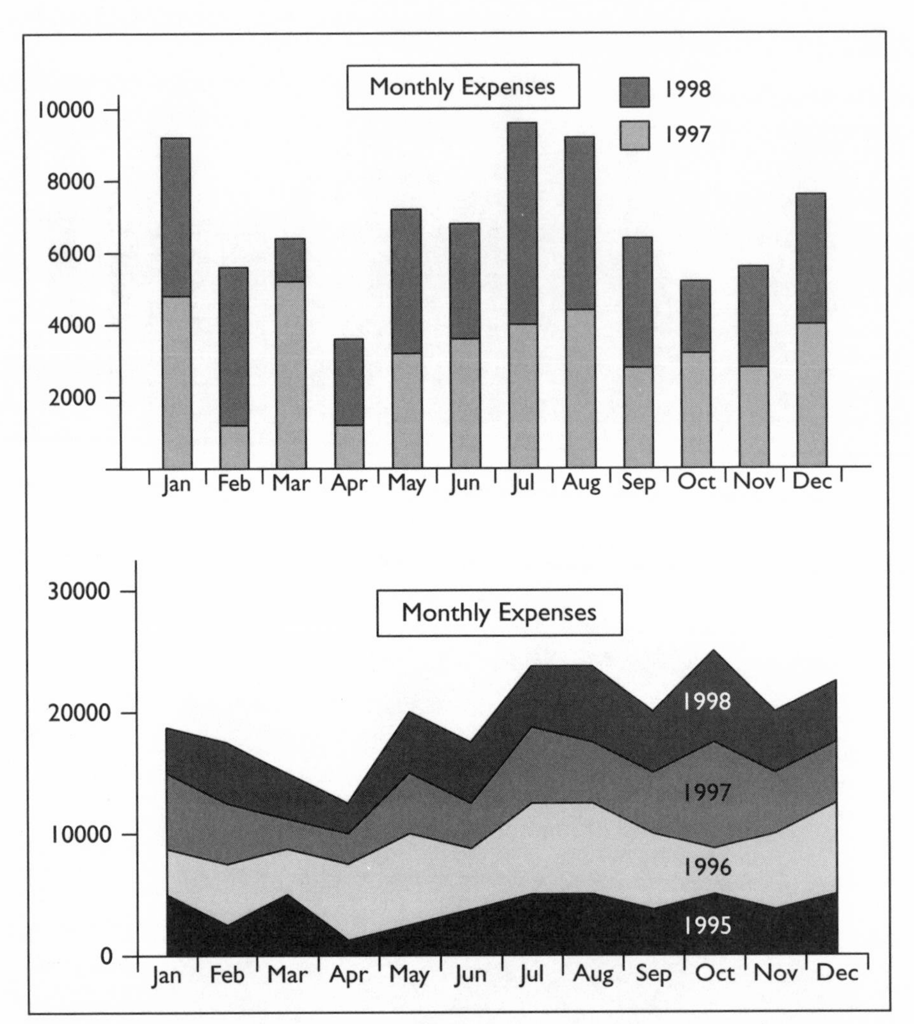

Figure 5-4. This shows a stacked bar graph at the top and a stacked area graph at the bottom. Both allow for easy comparison of data. In this case, the two figures add data from year to year, providing a clear representation of expenses in different years. The stacked area graph is especially useful for showing such data over three or more years.

Maybe it's best not to have any visuals at all. If so, fight the temptation to include them.

Finally, you should make sure your visuals are accurate. Check the numbers in your tables or graphs. Verify the labels

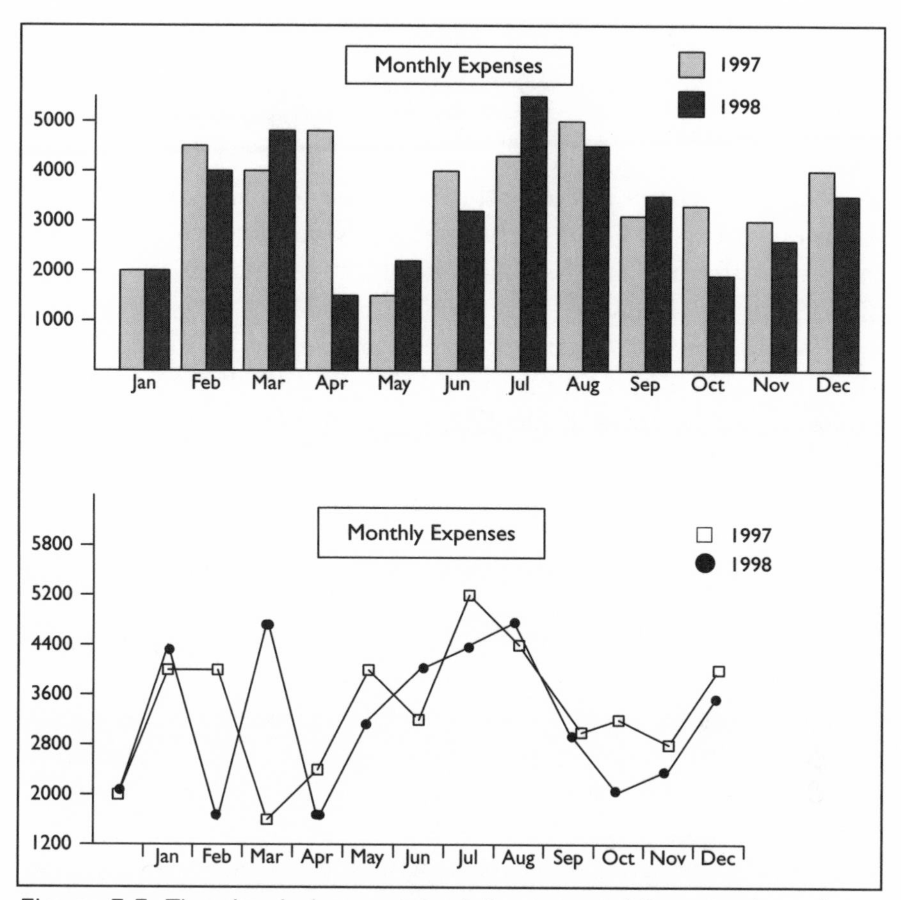

Figure 5-5. The simple bar graph at the top and line graph at the bottom do not add data but display two variables side by side for easy comparison. Both types of graphs can be effective for doing this.

and captions. Compare the information in your visuals with the information in your text. Remember to acknowledge the source of the data you use.

If you follow these three guidelines, that is that your visuals should be accessible, appropriate, and accurate, you'll make life easier for your readers and get the most impact out of your visuals and your text.

Where to Use Visuals

So, now that you know more about when and how to use visuals, the final issue is where to use them. That's considerably easier than the other issues.

Placing a visual at the end of a document suggests that it lacks importance. If you group several visuals at the end, it minimizes the impact that each would have individually if placed appropriately.

Placing a visual before you refer to it in the text suggests that the visual is more important than the text. It also can distract readers by raising questions. That technique is a great way to get readers thinking about an important problem or opportunity, for example, if that's part of your strategy.

Many experts believe it's most effective to place the visual immediately after the textual reference. That's what we've tried to do with the visuals in this book. It may not be easy, especially if your graphics are large or if you have a set of several graphics that require a whole page or facing pages.

Visuals are often an excellent way to present information, to gain and maintain reader interest. But don't underestimate the importance of visuals to provide relief from dense or otherwise demanding text. You want to make the most of that benefit for your readers by using white space and headings, as I'll discuss in a moment.

But first a warning about graphics: don't take them for granted. They can be powerful, but they can be dangerous.

Consider the table in Figure 5-6 in terms of the guidelines of accessibility, appropriateness, and accuracy. It would probably rate a 5 on a scale of 1 to 10. Why?

How could you improve this table? Did you notice that the numbers don't add up? That detail could undermine the credibility of your entire report. "Workshops" is actually a category under "Conferences," so the line should be indented to show that the $5,000 for "Workshops" is included in the $15,000 for "Conferences." Such a detail could confuse people reading the report, so the table would hurt rather than help your efforts.

Category	1998 actuals	1998 projections
Travel	40,000	45,000
Consulting	12,000	15,000
Conferences	15,000	18,000
Workshops	5,000	7,000
Visiting lecturers	30,000	35,000
Laboratories	43,000	48,000
Released time	220,000	230,000
Total	**360,000**	**391,000**

Figure 5-6. Wallace University research budget requests

Bottom Line on Visuals

Visuals can be a great way to emphasize important information, help you make your points, and draw your readers into your text. But inappropriate, inaccurate, or unattractive graphics may be worse than none at all.

Sometimes the most important visual touches are the easiest—white space and headings. Yet many managers underestimate the power of white space and headings or neglect to take advantage of that power.

White Space: Making the Most of Nothing

White space is a simple way to attract and maintain reader interest. I mentioned two ways to use white space in Chapter 4—indenting paragraphs and opening with a short paragraph—that provide entrance and exit ramps to ease readers into your document and suggest that it won't be difficult to read.

White space can be used in other ways. You can use ample margins to draw attention to your text. This technique can be particularly effective, for example, when you're listing items or quoting a source. A little white space between paragraphs allows each one to look like an easy-to-read unit. Long blocks of copy look gray and difficult to tackle. The block formats for letters (Chapter 4) show how a little white space makes a page of text look more inviting.

Here's a quick way to check out your use of white space. Just put a little distance between you and your text, maybe ten feet—at least enough so you can't read the words, so the text makes only a visual impact. What impression does it make? This test is an easy way to judge the first impression your text will have on the readers.

Headings: Signposts and Signals

Use headings to break up your text, to add a little black to some white space. Headings make it easier for your readers to access your text. That's a benefit you gain automatically, even if the headings were just words you chose at random. But using headings carefully offers other important benefits—both for you and your readers.

Your headings should show your readers the organization that you developed through your outline. Use level-one heads to mark the main sections, level-two heads to mark the subsections, level-three heads to mark parts within the subsections (if necessary), and even level-four heads (for smaller elements).

Headings help your readers understand and appreciate your organization and find what they want more easily. They also allow you to emphasize your text and mark transitions between topics. Finally, they have a sometimes critical advantage: they help keep you on track when writing a long text, so you're less likely to wander or to emphasize some sections while skimping on others.

Put a heading every two or three pages, but no more than seven level-two (minor) headings per level-one (major) heading. Mark the relative importance of your headings by using boldface, capitals, italics, and an appropriate point size.

Consider the four levels of headings shown in Figure 5-7 as an example.

MAJOR HEADING
(level-one: all caps, bold)

<u>Minor Heading</u>
(level-two: mixed case, bold, underlined)

<u>Subheading</u>
(level-three: mixed case, normal font, underlined)

Paragraph Headings
(level-four: indented, mixed case, normal font)

Figure 5-7. Four levels of headings

For another example, one that uses three levels of headings, look at this book: level-one heads are flush left, mixed case, bold, and larger; level-two heads are also flush left, mixed case, and bold, but not as large; and level-three heads lead into the following text, but in bold.

The following examples illustrate how headings help, even in a short text: the first version (Figure 5-8, page 80) uses no headings, and the second version (Figure 5-9, page 81) uses headings, white space, and visuals (bulleted lists).

Summary

You can put more power into your reports by using visuals, white space, and headings. Try to imagine your words from the perspective of your readers. Are your pages just blobs of gray, with all the graphic appeal of the telephone directory? Or do they grab and hold the interest of your readers, making it easier for them to follow you?

Many managers forget about the human factor when they write. They ignore, forget, or underestimate the power of visuals, white space, and headings. Any reasons they might have for this oversight don't seem very valid.

Make Your Writing Clear

The most valued writers in the organization write honestly. They transfer information so their audiences know it and feel it as they do. Granted, there are times you may want to be vague to leave room for debate and creativity. But much writing on the job is vague, undisciplined, and imprecise.

One of the problems that plagues writing on the job is "third-degree writing"—writing that's too general and doesn't honestly represent what's in the writer's mind. Example: The boss sends a memo saying that the situation in the office is getting serious. *Situation* isn't what the boss means. It's a third-degree word. Imagine a ladder—often called the ladder of abstraction in writing. The ladder has three steps. The bottom rung represents third-degree writing—situation in the office. The next rung is second-degree writing—dispute. The top rung represents first-degree writing—employee disagreement about smoking on the job.

The boss actually wanted to talk about the dispute employees are having over smoking on the job. But the boss called it a situation. The boss is a third-degree writer. First degree—disagreement about smoking, second degree—dispute, and third degree—situation.

If you want to convey a message, use concrete words. Avoid *office equipment* when you mean *computers* and *writing implement* when you mean *pen*. If you want people to understand you, be as specific as possible. Vague wording just about guarantees confusion. Well-intentioned managers who try to be democratic or non-directive often offer statements like "Joe, give this a top priority" or "Jill, I need this out as soon as possible." To avoid misunderstandings, specifically state when you need the job done.

Reprinted with permission from *Power-Packed Writing That Works* (1989).

Figure 5-8. Passage of writing without use of heads to guide readers. Compare this with the same passage just below on the next page.

Do they think headings take too much time to write? A few words in a heading can bring more value out of the hundreds of words that follow. Do they think white space wastes space? It's a good and easy investment. Do they think visuals are too difficult? Many computer programs make graphics simple.

MAKE YOUR WRITING CLEAR

The most valued writers in the organization write honestly. They transfer information so their audiences know it and feel it as they do. Granted, there are times you may want to be vague to leave room for debate and creativity. But much writing on the job is vague, undisciplined, and imprecise.

Third-Degree Writing

One of the problems that plagues writing on the job is "third-degree writing"—writing that's too general and doesn't honestly represent what's in the writer's mind. Example: The boss sends a memo saying that the situation in the office is getting serious. *Situation* isn't what the boss means. It's a third-degree word. Imagine a ladder—often called the ladder of abstraction in writing. The ladder has three steps:

- The bottom rung represents third-degree writing—situation in the office.
- The next rung is second-degree writing—dispute.
- The top rung represents first-degree writing—employee disagreement about smoking on the job.

The boss actually wanted to talk about the dispute employees are having over smoking on the job. But the boss called it a situation. The boss is a third-degree writer:

- First degree—disagreement about smoking
- Second degree—dispute
- Third degree—situation

Specific Words Tell It Like It Is

If you want to convey a message, use concrete words. Avoid office equipment when you mean computers and writing implement when you mean pen. If you want people to understand you, be as specific as possible.

Vague Wording Causes Problems

Vague wording just about guarantees confusion. Well-intentioned managers who try to be democratic or non-directive often offer statements like "Joe, give this a top priority" or "Jill, I need this out as soon as possible." To avoid misunderstandings, specifically state when you need the job done.

Reprinted with permission from *Power-Packed Writing That Works* (1989)

Figure 5-9. A passage of writing that includes headings to guide readers

Heading for Better Results

Headings should interest the reader and indicate the nature of the text to follow.

Whether you allow your headings to be funny or cute or keep them serious depends on the purposes of your report, the situation, and your organizational culture.

It may be more helpful to your readers to make the headings grammatically parallel—such as "Writing for Business," "Writing for the Reader," and "Preparing Your Text," rather than "Business Writing," "Writing for the Reader," and "Prepare Your Text." Parallelism may also be important if those headings will appear together in a table of contents.

It doesn't take a lot of effort or ingenuity to make use of visuals, white space, and headings. Most of the suggestions in this chapter rely on common sense and a basic understanding of psychology. You just need to care about appealing to your readers and helping them get more out of your writing.

Manager's Checklist for Chapter 5

- ❑ Make sure visuals support text, convey information, or direct action.
- ❑ Use the guidelines of accessibility, appropriateness, and accuracy when developing visuals.
- ❑ Know the advantages and disadvantages of the various types of visuals. Use the ones most appropriate for your purposes.
- ❑ Use white space generously to make your text easier to read.
- ❑ Put headings throughout your reports to break up your text, to make it easier for readers to get into it, to show how your report is organized, to help your readers find what they want more easily, to emphasize your text, and to mark transitions between topics.

How to Start and Stop

I gave a lecture one fall at Villanova University about using exciting introductions to gain attention. One student took my advice. As she started her speech, the other students began to fidget and stir uncomfortably in their seats. She started by describing how she and her boyfriend had experimented with many illegal drugs.

Several students felt embarrassed for her and glared at me as if I should stop her speech. She went on and on, providing intimate details of their drug use. Silence overwhelmed the classroom.

Finally she said, "That was the story of Laura Edward, a young woman with a difficult lifestyle to support." The class collectively gasped in relief. The speaker had certainly gotten our attention—a key ingredient to a good introduction.

Gaining Attention

What makes good sense for an oral presentation also makes good sense when you're writing. You need to interest your readers.

In Chapter 4 we discussed using a subject line in letters

and memos to focus the reader on your message. We also discussed formats that open your letters and memos by appealing to the reader. In this chapter we apply the same logic to the longer documents you write.

Most professional writers will advise you to start with a quotation, a story, an anecdote, an example, or even a rhetorical question to attract readers to your document. If that opening is appropriate to the purposes and contents of your document, you should then, with a little effort, keep those readers interested.

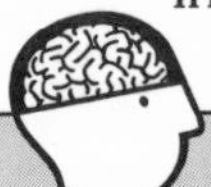

Use Anecdotes to Show Instead of Tell

Susan Perloff, a Philadelphia writer and writing teacher, advises using anecdotes to open your report. Anecdotes show something about your subject, which is usually more effective than just telling about it. They can make great leads because readers tend to remember them.

Start Strong, Finish Strong

Make your introductions and conclusions strong. This emphasis is important not just because you want to start strong and finish strong but because the reader may only read the beginning and the end. Readers who are pressed for time will read only an executive summary, an introduction, or a conclusion. For this reason, your introduction and conclusion should provide critical information. Of course, if you make them really interesting, your introduction and conclusion may also motivate some people to read the entire document.

Degrees of Attention

You must write to attract attention—unless you intend your document to go immediately into a file somewhere. (In many businesses, writing "to record for posterity" seems to be a legitimate purpose. If you just want to fill paper with words for a file, you really don't need to be concerned about attention.)

How important is it to attract the attention of your readers? That depends primarily on three factors:

- Person and/or position
- Subject
- Purpose

Who are you and what position do you hold? Let's face it: we're not all equal. Everybody in a company may be interested in whatever the president writes, but perhaps less interested in a report from the assistant to the associate manager of office supplies.

What are you writing about? Certain topics will have greater natural interest for your readers than will others. Fortunately, with a little work you can change the natural level of interest of almost any subject.

Why are you writing? Your central purpose influences the degree to which it's important to attract your readers. In Chapter 2, we discussed four basic purposes for writing: to inform, to persuade, to instruct, and to record. Things get somewhat more complex when we also consider the reasons behind the purposes.

If you're writing to inform or to instruct, is it because somebody else wants the information or instruction? Or is it because you want to provide it? In other words, will your readers feel motivated because of their need for the information or instruction, or will it be necessary to stir their interest?

If you're writing to record, your purpose may not be that complicated. You may simply want to put something down in writing for posterity, without worrying about whether anybody will read your writing. But if you want to impress the boss, you'll want to make sure she actually reads the report.

If you're writing to persuade, then it's crucial to interest your readers. You need to move them with your words, so you've got to draw them into those words or your correspondence is a big waste of effort.

So, just as it's essential to know your single purpose before you write, as we covered in Chapter 2, it's essential to think about how much you need to interest your readers and which way might work most effectively.

Effective Openings

Introductions are crucial, for the reasons we've just reviewed. They're also the most difficult part of writing for most people. You make the job easier when you've thought about your purpose and your particular needs to arouse interest—and when you've organized and strategized with an outline.

But that doesn't mean that writing an introduction will be a breeze. It still could be tough. That's why I suggest the CPO method for writing introductions, as recommended by Michael L. Keene, author of *Effective Professional Writing* (1993) and co-author of *A Short Guide to Business Writing* (1995).

CPO is a mnemonic device for remembering the three essentials of a good introduction:

- **Context**—Establish the circumstances for writing, the setting for your words.
- **Purpose**—Give your reasons for writing.
- **Organization**—Indicate what you'll be doing in your report.

It's a basic approach that you can keep simple or use creatively. Let's look at four introductions to see how well they open with context, purpose, and organization.

Example Introductions

Five Ways to Write Better

This section suggests five techniques that will help you write better. It doesn't matter if you have a Ph.D. in English or if you struggled with language all through school. You just need to want to write better—and be serious about committing to that goal.

Use all five techniques together or use them separately, depending on your needs. They can help you write more effectively, whatever your business needs may be.

- *Context:* The context is improving writing.
- *Purpose:* The purpose seems clear: to provide instructions for writing better.
- *Organization:* The organization is indicated in the first sentence: this section will present five techniques.

How to Pack a Punch with Your Proposals

You run into the boss one evening as you're leaving the building. You happen to mention an idea that's been on your mind for a while, a way to expand your market at very little cost.

She loves your idea. In fact, she asks you to write up a proposal as soon as possible. It's a great opportunity—and a frightening experience.

We've all been there. We know how many good ideas fall apart when they make it into a proposal. A manager who cannot describe an idea completely yet concisely, with precision and with passion, cannot deliver the proposal punch that can put that idea into practice.

- *Context:* The context is conveyed through a common experience.
- *Purpose:* The purpose is to help the readers express their ideas in proposals that generate action.
- *Organization:* There's no indication of the organization.

Twenty-First-Century Public Relations (PR)

To prepare for this presentation, I consulted with four practitioners in the fields of government affairs, health care, telecommunications, and investor relations. I asked them about the following issues:

- Technology and its impact on the public relations profession in the next millennium.
- Globalization and how it would affect PR roles.
- Media relations and how they would differ in the next century.
- Spinning and the law, particularly the role of PR in using the media to form opinions before a legal battle is decided in court.

I'll cover each area and its impact.

- *Context:* The context is provided primarily through the reference to practitioners—an attempt to create shared experience.
- *Purpose:* The purpose seems implicit, to provide an overview of changes in PR.
- *Organization:* The organization seems clear—looking at four separate areas, with bullets to emphasize that organization.

Investing in the Future Towing Business of a Released Prisoner

My name is Harry Harm, and I am presently confined in the Virginia Department of Corrections. This letter is being sent to you with the hopes of obtaining your assistance in my becoming an entrepreneur after my release from prison. During the past fourteen years of my life, I have been confined in various penal facilities.

Yes, I took a wrong turn in life, which resulted in my placement in a more structured environment. Being confined is not something one should take lightly, and every man should value his freedom to the fullest. For someone my age (39), getting out and staying out is mandatory. One of the means by which this can be accomplished is by the creation of my own business.

- *Context:* The context is entirely for the writer; there's no shared context. In fact, from the title on, the writer stresses the differences between his context and the context of the reader.
- *Purpose:* The purpose comes in the last line of the introduction—to stay out of prison by enlisting investors to help the writer start a business.
- *Organization:* There's no indication of the organization.

Make the Opening Attractive

The CPO method ensures that you'll provide the three essentials in your introduction and improve your chances of drawing your readers into the body of your document. As the four examples have shown, you can indicate context, purpose, and organization in various ways. How you do this depends primarily on your readers, your subject, your work culture, and the situation.

When writing your introduction, remember to keep the few first paragraphs short, allowing the reader an exit, or "off-ramp"—a way to get out quickly if he or she doesn't find the start interesting.

Holding Your Readers

Once you've hooked your readers, you must keep them interested. Readers become frustrated when lured into a document by promises of interest that quickly prove to be empty. If you start with an anecdote, for example, then continue with human touches throughout the document. If you indicate context, purpose, and organization in the introduction, then build on that context, pursue that purpose, and follow that organization.

We discussed in Chapter 5 the use of headings as signposts and signals to guide readers through your text and hold their interest. You can also use transitions to hold and guide readers.

Appropriate transitions can improve the flow of your sentences and paragraphs and make your text easier to read and understand. Here's a list of some words and phrases commonly used as transitions:

- **Go ahead:** and, moreover, furthermore, also, for instance
- **New idea:** thus, so, and so, therefore, consequently, accordingly
- **Summary:** as a result, at last, finally, in conclusion, in short
- **Change idea:** but, yet, nevertheless, otherwise, although, despite, however, conversely
- **Link cause and effect:** that caused, as a result, that produced, consequently
- **Refer:** they, these, though, not one, all but two, without exception
- **Restrict and qualify:** provided, in case, if, lest, when, occasionally, even if, never

Source: *communication briefings*

In addition to transitional words and phrases, there are other ways to guide your readers through the text. A very good way to link sections of a text is simply to use a sentence or two to sum up where you've been and to give an idea of where you're going.

Take the sentence that began the previous paragraph, for

example: the first part—"In addition to transitional words and phrases"—summed up the preceding section, while the second part—"there are other ways to guide your readers through the text"—told what was coming in the next section. This sort of transition can be straightforward, subtle, or striking, as appropriate to your circumstances and needs. We could have begun the previous paragraph with something like "Wow! That's an impressive list of transitions. But what if I don't want to use just a word or a phrase?"

Remember that transitions serve as signposts and signals. They can only help ease the connection from one sentence to the next, from one paragraph to the next, and from one section to the next. If you find it difficult to provide transitions for your readers, go back to your outline, because you may have some problems with your organization.

> CAUTION!
>
> **No Magic Cure**
>
> Transitions can make good writing better, but they're not a panacea.
>
> Don't insert transitional words and phrases throughout your text in hopes of making it seem more coherent. Consequently, it's a bad idea.
>
> (How did you feel when you read that last sentence, after the transition word "consequently" announced a cause-and-effect relationship or a new idea, but you found just a statement of judgment? Disappointed? Confused?)

Transitions will take your readers from your interesting introduction through every section of your text. Transitions work with the interest you generated at the start—and they should help you build that interest toward a powerful conclusion. (How was that paragraph as a transition into the final section of this chapter?)

Ending Strong

Most readers remember best what they read last. What a great opportunity to move your reader in the direction you choose! But many documents end as if the writer simply lost interest or felt confident that the job was already done.

Your conclusion should be as strong as your introduction.

You write the introduction to generate interest; you write the conclusion to sustain that interest.

Wrapping It All Up

Your conclusion should do the following:

- Briefly summarize the document.
- Repeat your major points for emphasis.
- Stress once again the importance of the topic.
- Provide recommendations, if appropriate to your purposes.

When should you use recommendations? That depends on your purposes—not only your central purpose for writing the document, but also any secondary purposes. If you're writing simply to inform or to record, you may not need recommendations unless you're citing ways to put the information to use, for example, or outlining possible consequences of the actions you're recording. If your purpose is to instruct, you may suggest ways to practice those instructions and to put them to work. If you're writing to persuade, your must offer recommendations. We now focus on how you can do that most effectively.

Always Recommend an Action

The most common mistake in writing to persuade is to stir up the readers and then leave them hanging with nothing specific to do. In other words, the writer does all the work, but isn't likely to gain the desired results.

So you've done your best to persuade your readers. Now tell them what you'd like them to do.

This may be difficult: most of us don't want to seem forceful—or maybe we're afraid of failing. But think about this: if you've persuaded your readers, if you've moved them to want to do something, then you're letting them down if you don't recommend action. In a sense, you're no longer the writer working alone.

How can you present your recommendations? You can do it very simply, by just writing from your readers' perspective. You start with a heading—"Recommendations" or "What We

Must Do" or whatever wording seems most appropriate to your message and your readers. You signal your transition: "Now, what can we do to solve this problem?" or "This situation is growing worse, and we can no longer continue to do nothing." Then you present your recommendation(s): "I see two approaches to solving this problem" or "We must immediately..." or "I strongly urge management to take the following action..."

Use specific language to focus the interest and energy of your readers. If you can't be specific, you may need to think a little more about your reasons for writing.

End as You Began

You may want to simply end with your recommendations, to pass the torch to your readers. But sometimes that sort of ending may seem too abrupt, a little incomplete.

You can often end very effectively by returning to your beginning. People generally like closure, so starting and ending in the same way completes the process for them. If you began with a quote, end with a quote. If you began with a story, you might be able to return to that story. What works best is to begin with the start of a story, then close with the end of that story, keeping the reader in suspense throughout your document. But move from the story into the body of your document gracefully, to keep the readers with you so they won't simply go straight to the conclusion to find out what happens.

Let's end this chapter with a look at the conclusion of the document on twenty-first-century public relations that we started in the section on introductions.

> So, in summary, we can expect much change in the areas of technology, globalization, media relations, and pretrial opinion forming. Having surveyed various practitioners, I would like to conclude with their recommendations. First, watch for an increasing number of female practitioners in leadership roles. Look for the return of small, specialized public relations agencies that serve a handful of clients, rather than thirty. Become educated as a "generalized specialist," someone who holds a general knowledge of public relations plus understands a particular industry well, e.g., health care.
>
> The experts predict the need for more practitioners with international knowledge. And last, we need to become committed to lifetime education and to increase our knowledge in other related fields such as psychology, economics, and philosophy. The future looks bright for public relations practitioners in the next millennium.

How would you rate this conclusion? It starts with a transition, "So, in summary, ..." It refers back to the four areas listed in the introduction and covered in the body. But it doesn't reemphasize the major points. The summary contains recommendations, but they're not very forceful: "watch for" and "look for" are calls to observation, not action, whereas "become educated" and "we need to become committed" are too vague.

As you read the reports that cross your desk, check out the introductions and the conclusions. How well do they work? Why? How could you improve them? What transitions do the writers provide for the readers?

Manager's Checklist for Chapter 6

- ❑ Put context, purpose, and organization in your introductions.
- ❑ Attract the attention of your readers with a story, anecdote, question, example, or startling statistic.
- ❑ Maintain interest by using transitions to guide readers.
- ❑ Conclude by summarizing, repeating main points, and making appropriate recommendations for action.

Zero-Based Writing

In Chapters 5 and 6 I focused on writing longer documents, such as reports, for the purpose of informing and/or persuading the reader. In this chapter, I focus on writing to provide instructions, to tell the reader how to do something.

Move from Assumptions to Perspective

Many managers make a critical error when writing instructions—they tend to make too many assumptions about what the reader knows. Especially when we write about nontechnical subjects, we assume a certain knowledge and a certain level of experience.

That's why analyzing your readers, as we discussed in Chapter 1, will help you connect with what they understand. But making that connection is not easy to do, because it requires that we open up our minds to take on the perspective of our readers.

This little exercise will help you realize how much we take for granted.

Starting from Zero

Pretend a family completely unfamiliar with what you do, maybe from a Brazilian rainforest culture, is visiting you, and the members of this family have never had any contact with the United States or European countries. They're hungry, so you suggest that they make some peanut butter and jelly sandwiches, a common food in the United States. Write a set of instructions for this family, assuming they know nothing about peanut butter, jelly, or bread, and tell them how to make a sandwich. Remember: no one in this family has the knowledge you have. They're starting from zero.

> **An Early Lesson**
>
> *For Example*
>
> A friend told me the following story.
>
> His six-year-old son loved to tell him about everything that was happening in his life. But his narratives were always very confusing. So my friend just started suggesting, every time his son began a story, "Tell me as if I wasn't there."
>
> That simple suggestion made a big difference. His son began thinking outside his own perspective, so he became better at telling his stories. It turned out to be a very important lesson for that six-year-old: think in terms of the other person.

You might come up with instructions like the following.

Follow me to the kitchen, where we make our sandwiches. You will find various food substances stored in this kitchen. Ask a person for four items: a loaf of bread, a jar of peanut butter, a jar of jelly, and a knife. The bread, peanut butter, and jelly will each have a label stating the name of the product. The knife is an eating implement used to place the peanut butter and jelly on the bread. Now we're ready to make sandwiches.

1. Take two pieces of bread, called slices, from the whole bread, called a loaf.
2. Place them on a horizontal surface.
3. Open the jar of peanut butter by twisting off the lid, or top.

4. Take the knife, or eating implement, and place the thinner end into the jar of peanut butter.
5. Put an amount of peanut butter on the knife and transfer it to one of the slices of bread.
6. Spread the peanut butter on the bread with the knife.
7. Open the jar of jelly.
8. In the same manner, place the thinner end of the knife into the jar of jelly.
9. Put an amount of jelly on the knife and transfer it to the other slice of bread.
10. Take one slice of bread and put it on top of the other slice, with the peanut butter or jelly side down.
11. Prepare to enjoy your first sandwich.

The Trouble with the Unknowing

That was easy, wasn't it? But my instructions have a few problems.

Let's start with the question of vocabulary. I used the word "kitchen," which these people might not understand. I cheated a little, because I told the visitors to ask somebody for the food and the knife. That simplified the task for everybody. I then described how to recognize the peanut butter, jelly, and bread, but I described the knife only in terms of our intended use of it—"an eating implement used to place the peanut butter and jelly on the bread." How would this family understand what implement we mean, since we haven't yet explained how to use the knife?

The vocabulary problems continue in my instructions. Why give this family technical terms like "slices" and "loaf" if they only need to make sandwiches and they can do so with the less technical words "pieces" and "whole"? Then, in instruction 4, I tell them to take "the knife, or eating implement." But I've already described the knife, so I only complicate the instructions here by referring to it as an "eating implement."

These instructions have other problems. For example, look at instructions 5 and 9, where I tell the family to put "an amount" of peanut butter and jelly on the knife. I should probably be more precise, because they don't know at this point what they're going to be doing with the slices. (Sure, we all know what a sandwich is, but do they?) Then I begin instruction 8 with the phrase "in the same manner." What does that mean here? Is this phrase necessary? The instruction seems just as good without it.

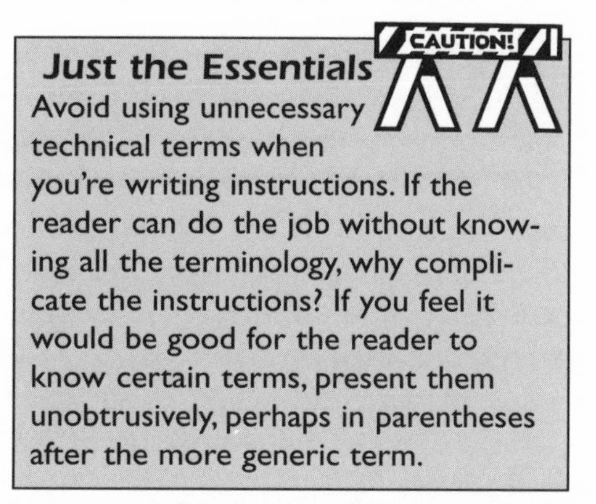

Just the Essentials

Avoid using unnecessary technical terms when you're writing instructions. If the reader can do the job without knowing all the terminology, why complicate the instructions? If you feel it would be good for the reader to know certain terms, present them unobtrusively, perhaps in parentheses after the more generic term.

Now consider instruction 10: "Take one slice of bread and put it on top of the other slice, with the peanut butter or jelly side down." That's clear to us, because we've been making sandwiches since we were kids. But which slice is it "with the peanut butter or jelly side down"? Did I indicate that it was the slice they should be picking up? They could follow those instructions and end up with the peanut butter and jelly sides facing out—a real mess. I should have been more exact in placing our modifying phrase—"and put it peanut butter or jelly side down on top of the other slice"—or specified "so that the peanut butter and jelly come together."

Finally, the last instruction is to "prepare to enjoy your first sandwich." Sure, it's just a casual comment, but how would the family know that? Maybe they would be confused by the command "prepare." Prepare how? Why not just give them the basic instruction "eat"?

Even such a simple task as making a sandwich can be complicated for those who are unfamiliar with the process and its vocabulary and techniques. Unfortunately, our instructions can make a task even more complicated if we make too many assumptions and if we don't consider the process from the perspective of those who will be following our instructions.

We could make it easier for this family to follow those instructions if we included visuals or diagrams with our description of the process. A drawing of a knife would save us a lot of words, because we wouldn't have to explain how to distinguish a knife from other implements that might be used for eating—just in case a human wasn't in the kitchen to help. In fact, we should provide drawings of several knives, so they aren't lost if they can't find the specific knife that we might draw. We have an even bigger problem if these people are unfamiliar with the technique of spreading!

It's Not Easy to Write for the Uniformed

This example might seem ridiculous, but it makes us aware of the many difficulties involved in writing instructions. We need to forget all our assumptions, consider the process from the perspective of the people who will be using our instructions, and cover all the basics.

We don't work with people who are completely uninformed. But we certainly feel at times like we're dealing with different worlds. Take a computer manual as an example. How much new or different vocabulary is used with no explanation, except perhaps a reference to a glossary that's even more confusing? How many concepts did the authors of that manual assume that we would understand? It's not easy to give instructions for a process—especially when we're very familiar with that process.

Master Process Writing

Process writing sounds technical and difficult. But it's simply a matter of telling your readers about something. Most process writing includes a descriptive overview and a set of instructions.

User manuals are examples of process writing. They usually start with a description of a process, such as a computer program or a VCR, and then provide detailed instructions for how to use the product. You could easily describe many processes you do each day at work, such as accessing e-mail messages or making coffee.

Remember: when you describe a process, make sure others could use your description or set of instructions to work through the process without you. Avoid assuming anything about your readers' knowledge or experience. Consider the process from their perspective. Then describe the process in a logical order and provide the necessary instructions step-by-step.

Now, let's break our process writing down into its two main parts: descriptions and instructions. We'll discuss descriptions first, then instructions.

Write Clear, Practical Descriptions

What does it mean for a description to be "practical"? The description should cover the essentials, everything the reader needs to know, and little or nothing else. Take our family, for example. They needed to learn how to make sandwiches. To do that, they didn't need to know who invented sandwiches or how we make peanut butter or the nutritional composition of a peanut butter and jelly sandwich. That information might have made the activity more interesting, but it would have complicated our instructions and their learning experience.

It's difficult to describe things, especially when we don't think of describing as difficult. Where do you start? Why? What do you include? What should you leave out of the description? These are the fundamental ques-

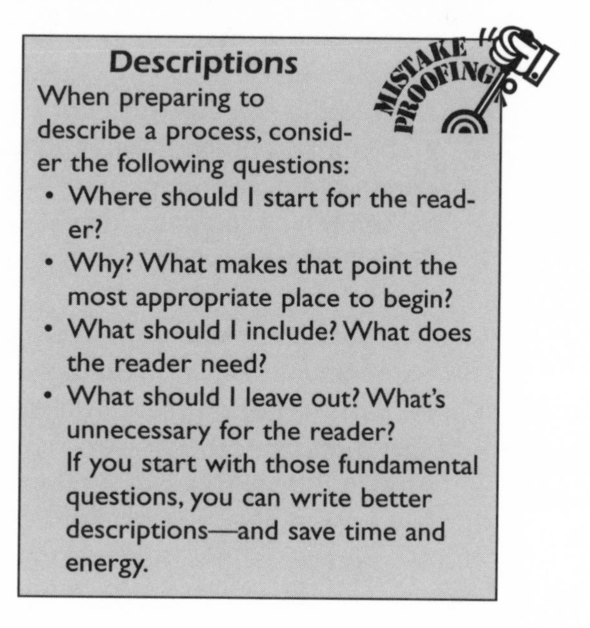

Descriptions

When preparing to describe a process, consider the following questions:

- Where should I start for the reader?
- Why? What makes that point the most appropriate place to begin?
- What should I include? What does the reader need?
- What should I leave out? What's unnecessary for the reader?

If you start with those fundamental questions, you can write better descriptions—and save time and energy.

tions we need to answer when writing a description of a process.

To describe a process, we must know and understand it. Then we should approach it as if it were all new to us, as our readers may be approaching that process.

We've acknowledged that it's difficult to take off our perspective and put on the perspective of an outsider. But we can become better at changing perspectives with a little effort. We can use various techniques to stimulate our thinking about describing what we know.

Business writing author Michael Keene recommends ten or twelve techniques to create descriptions. The following exercise uses some of those techniques. We'll run through this exercise to come up with various ways to describe *spring*—something very familiar to us all.

Write a brief description of spring based on each guideline.

- **Formal dictionary definition:** One of four seasons, spring follows winter and leads into summer; it is the season of rebirth.
- **Accumulation of detail:** In spring the birds return north, flowers and grasses bloom, trees flower and become green; various animals give birth during this season.
- **Process:** One of the two times of year when the sun crosses the celestial equator and when the day and night are approximately equal in length.
- **Elimination:** Spring is unlike winter, when it becomes colder. It is unlike summer, which is drier and hotter. It is unlike autumn, when leaves fall off the trees in preparation for winter.
- **Compare/contrast:** Spring is most like autumn, which is a transitional season leading into a more extreme weather pattern, winter. Spring is unlike the extreme seasons of winter and summer.
- **Analogy:** Spring is like a rebirth, when everything comes alive again, like a reawakening or a resurrection.

This exercise shows how we can approach a subject in a half-dozen ways to write a description. How we decide to approach our subject depends on the readers for whom we intend our description. Which approach would be more appropriate for someone starting from zero?

Pachyderm Problem

For Example

You may remember that poem about the six blind men who were trying to describe an experience new to them—an elephant. One described it in terms of its trunk, another in terms of its tail, while the others seized upon a leg and its skin and so forth.

Which one is right? All of them. But all of them are also wrong.

It's a matter of perspective. Find the perspective that's most useful to your readers and their need for your description. That's where you begin.

Sometimes you may be able to jump right into describing a process. But it may help to start with an exercise like the one above. Once you're able to approach a subject in various ways, you can more easily write an overview or an introduction to any process.

What should you include in your description? Whatever it takes to set the stage for the instructions to follow. You might give the reasons for the process, explaining how it fits into the larger work system. Part of that context might be any preparation required for the process. You might tell how the process works in terms of inputs, outputs, mechanisms, or whatever. Depending on what your instructions will be guiding your reader to do, you might go through the process step by step.

You might also include graphics. But don't be misled by that saying about a picture being worth a thousand words. Don't expect a graphic to do the work of words; graphics should be supplements, not substitutes, for words. Describe the particular aspect, then use the graphic to illustrate what you mean.

If you know the process and you know your readers, you've solved the question of perspective. Write your description—and then put it to the test. Read it as your readers would. Is there any aspect that could possibly confuse them? Is there anything missing that they might need to know? What questions might come to mind? If your description is confusing or incomplete, revise it. Remember: a good description makes it a lot easier for you to write instructions—and a lot easier for your readers to understand and follow them.

Next we move on to writing complete, precise instructions.

Write Simple Instructions

Instructions should first be simple, but should also be complete, precise, direct, and organized. But what does that mean in practice? The following guidelines should help you write better instructions.

- **Complete.** Include every step, together with any necessary information such as standards or explanations.
- **Precise.** Specify as exactly as necessary the tools to use, the amounts to measure, and all other aspects of each activity.
- **Simple.** State in short sentences only what readers need to do or to know. It's best to use imperatives: readers will understand them more easily—and using imperatives will force you to focus on actions and use the active voice
- **Direct.** Keep your focus on the activity and your readers. What must the readers do, and what instructions and information must you provide in order for them to do it?
- **Organized.** Put all the steps in the proper order. If there are two actions that the readers must perform simultaneously, include them both in the same instruction, preceded by some indication, such as "Do both of the following actions at the same time."

Think back to our instructions for the rainforest family. How did those instructions rate according to these five criteria? Our instructions were fairly complete. They were not very precise—we didn't specify the amount of peanut butter or how the slices were to come together. Our instructions were simple, although we didn't need to use such terms as "slice" and "loaf." We could also have split instructions 5 and 9 into two actions. Our instructions could have been more direct: for example, we should have said simply, "Twist off the top of the jar of peanut butter." Finally, our instructions were organized well—not surprising for such a simple, linear task.

Layout

Now that we've discussed what we should include in our instructions, we will consider the format. As we stressed in Chapter 5, appearances can really help or hurt our writing.

Most readers find compact paragraphs of italicized instructions tedious. They tend to toss them aside to try the process themselves. All the effort that went into writing those instructions is wasted because those words were formatted as dense paragraphs in italics.

When you format instructions, use bullet points or numbers to mark each of the steps. If you're providing instructions for several sets of activities, it might help your readers if you supplied headings to make the organization stand out visually. Use italics only to emphasize occasional words. Keep each instruction short—which should be easy if the instruction is simple and direct.

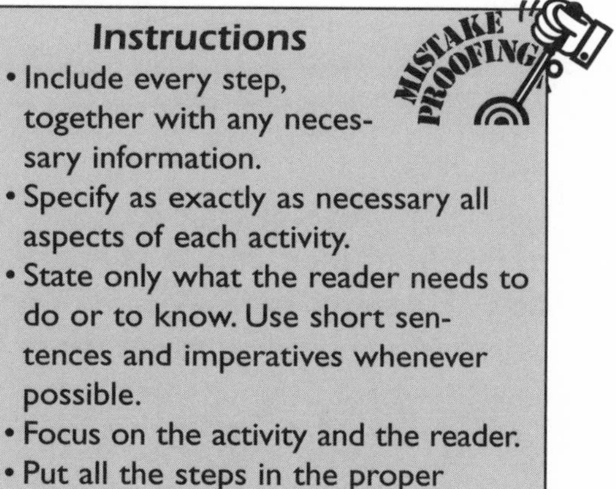

Instructions

- Include every step, together with any necessary information.
- Specify as exactly as necessary all aspects of each activity.
- State only what the reader needs to do or to know. Use short sentences and imperatives whenever possible.
- Focus on the activity and the reader.
- Put all the steps in the proper order.

Writing instructions requires logical, sequential thinking. It then requires a very critical eye. Read through your instructions aloud, slowly. Act them out—or at least mentally picture each action. Replace any long phrases with shorter ones. Move verbs to the beginning of the sentence, as imperatives whenever possible. Eliminate every unnecessary word. Reword negative commands in positive terms; if you're explaining how to put paper into the copier, for example, "don't try stacking more than 150 sheets" is not as good as "stack up to 150 sheets."

Smart Managing

It's Only Logical

We don't all think in the same way. You know that—so let your instructions show that you know it.

According to a study by the University of Rochester in 1992, following directions is gender based. Men drive to a destination by counting mileage or going north, whereas women use landmarks like gas stations. University researchers tested male and female rats in a maze and found that females depended on landmarks and males depended on vectors or direction.

That's just one example of differences in the ways we think. Don't assume that anybody reading your instructions is going to think just like you. You could be very surprised!

We've advised you throughout this chapter to think from the perspective of your readers. We also want to remind you to feel from their perspective as well. Remember the frustration of assembling a bicycle or changing a cartridge in a laser printer while reading instructions that just didn't make sense? Use those experiences to guide you in writing instructions that reflect concern for the readers.

Put It All Together

We began this chapter with a family completely new to the United States to show how difficult it could be to write good descriptions. It seems appropriate to close the chapter with a more down-to-earth example involving an automated teller machine—an ordinary process in our lives.

How to Get Money the Easy Way

Our purpose: to explain to someone who has never used an automated teller machine (ATM) how to withdraw $100 from her checking account.

Withdrawing Money from an ATM

[description]

In the 1980s, banks devised a way to make it easier to make deposits and withdrawals—the ATM. These money access machines provide convenient, twenty-four-hour access to your accounts. The bank issues you an ATM card and you select a four-digit access number, or code, which you share with the bank. With this card, you can access your money through a number of ATM machines that are located not only at banks, but also at convenience stores, airports, casinos, and so on.

Some ATM machines are equipped with braille so that blind users can access their money as well. Some people fear that taking money out in public areas can produce dangerous results. Many ATMs today use cameras to photograph any such robberies or incidents.

You may conduct various transactions on an ATM, including withdrawals, deposits, and transfers. You can also request an account balance. Read on for exact instructions on how you can use these convenient machines.

[instructions]

First, decide whether you will drive or walk up to an ATM. Once you've approached the machine, follow these easy instructions.

1. Insert your card in the indicated manner (many times the machine will tell you to insert your card with the black stripe down and to the right.) If you don't enter your card correctly, it will come back to you.
2. After you enter your card, the machine will prompt you for your four-digit code. Put in the code and press Enter. If you make a mistake, press the Clear key and start again. CAUTION: If you enter the wrong code more than twice, the machine may take, or "eat," your card.
3. Once the machine approves your code, it will ask you what transaction you wish to perform. Indicate by pressing the identified key that you would like to make a withdrawal and press the Enter key.
4. The machine will then ask you from which account you would like to withdraw this money. Answer your primary checking account and press Enter.
5. The next screen will typically ask for how much money you would like. Often it will list different amounts and you may select one and press Enter. Note: the machine doesn't give coins and often will give only $20 bills.
6. Indicate that you would like $100 and press Enter. The machine will tell you to wait.
7. The money should be issued through a slot in the machine and you'll probably receive five $20 bills.

8. The machine will then ask if you would like to do another transaction. Indicate "No."
9. The machine will issue a receipt, which usually includes the amount of your withdrawal and the amount remaining in your account.
10. Take your receipt and don't forget to remove your card!

Evaluation of the Description and the Instructions

What do you think of this example of process writing? Is the description clear and practical? Are the instructions complete, precise, simple, direct, and organized? Bonus question: Where might it be useful to supplement the text with visuals?

The description begins well with a historical overview and an explanation. In the second paragraph, however, it begins to wander. It's nice that ATMs are accessible to people with visual problems, but does our reader need to know that? Why mention the security cameras? That's not information she'll need to use. If we're concerned about theft, we should include a final instruction not to show or count the money in public.

Then the description becomes less clear. The wording "various transactions... including" suggests that there are more than three transactions—not including requesting an account balance, which is certainly not a feature of all ATMs. The description would be more useful if it specified the possible transactions.

The instructions that follow that description don't rate very high either. We'll touch on a few of the problems and let you find the others.

The instructions seem fairly complete—although the first sentence would be better phrased as an instruction: "Choose an ATM." (Mention of the two basic types belongs in the description, not in the instructions.) Also, what if the reader chooses an ATM that's in a locked enclosure?

The instructions may be as precise as we could make them, given the variety of ATMs. In fact, they may be too precise: instruction 1 assumes that the swipe mechanism is always horizontal, never vertical. But when the instructions refer to information that the ATM will provide, the wording alternates between "machine" (vague) and "screen" (precise). We could also tighten loose wording here and there, as in instructions 2 and 6, for example, where "put in the code" and "indicate" might be better phrased as "press the keys to indicate."

The instructions seem simple, but only if the reader is familiar with ATMs. What happens if, as suggested in instruction 1, her card comes back? What happens if the machine takes her card? The instructions give the reader a second way to refer to that possibility ("eat"), but that knowledge won't help her react to that unexpected event. Also, why tell the reader, in instruction 7, "You'll probably receive five $20 bills." The denominations certainly matter less than knowing what to do if she doesn't receive the correct amount!

The instructions are not as direct as we should make them, either; a critical revision could improve the wording. To take only one example, instruction 3—"indicate by pressing the identified (?) key that you would like to make a withdrawal"—could be phrased more directly as "press the key marked Withdraw."

The organization seems appropriate—although it might be good to include some instructions about what to do in case of failures, such as if the card fails to register.

The Final Test

That's a quick evaluation of our description and instructions. But what really matters, of course, is how well they work. The final test when you write instructions should always be to give them to an ideal reader, somebody typical of the people

> **Tips**
>
> MISTAKE PROOFING!
>
> Write like you're starting from zero. Assess the knowledge and experience of your readers. Don't assume anything. Think from your readers' perspective. Provide clear, practical descriptions and complete, precise, simple, direct, and organized instructions. Then, test out those instructions—and be critical of your efforts.

for whom you wrote those instructions.

In this case, if the woman in question who has never used an ATM succeeds in withdrawing $100, your instructions have worked. On the other hand, if she can't even get her card in, you may need to review your instructions.

Manager's Checklist for Chapter 7

- ❑ Never assume that the reader knows what you know—or thinks the way you think.
- ❑ Write clear, complete descriptions to provide an appropriate, useful context for your instructions.
- ❑ Write complete, precise, simple, direct, and organized instructions.
- ❑ Use visuals appropriately to supplement your words—but never to supplant your words.
- ❑ Indicate caution or trouble spots that the reader may encounter—and give directions for dealing with those possibilities.
- ❑ Test out user manuals on someone unfamiliar with the process.

Dealing with Tough Situations

One of your employees has been letting her performance decline for the past six months, despite warnings. Her coworkers like her, but she's hurting your department with her bad attitude. Now you've gotten evidence that she's been running her own little business on company time.

Two of your primary regional distributors have written letters to you complaining that shipments have been arriving late in recent months. Even worse, some lots have been short and some of the products are damaged.

Your department is going through a financial slump. You've been forced to defer bonuses and raises for several employees. Now you emerge from a management meeting with orders to lay off three employees, at least temporarily.

A reporter from the local newspaper has written several articles about problems in the business community. He's accused your company of allowing discrimination and harassment. You suspect that one of his sources is a former employee in your department who recently quit under pressure and threatened to get revenge.

As you read this list of scenarios, maybe you feel your heart beating a little faster, your temperature rising slightly, and a lump forming in your stomach. If so, then this chapter is for you.

Few managers are comfortable dealing with tough situations. For most of us, that chore can be less difficult if we're confident that we're handling the communication as well as possible. That's our focus in this chapter—to take you beyond our brief consideration of tough situations in Chapter 4.

Breaking Bad News

Sometimes things are going well, or at least it seems that way, when a problem suddenly comes up. Now you've got to break the bad news. There may not be any good way to deliver bad news, but it can be easier if you remember these two points:

- Be positive.
- Give reasons.

Phrase the bad news in positive terms. At the very least, begin positively. There are two good reasons for doing so. First, negative words will naturally annoy the person reading your message. Second, research shows that it takes the mind longer to understand a negative statement than a positive one.

Strategy for Delivering Bad News

In Chapter 4 we discussed delivering bad news in letters or memos using the following format:

Paragraph 1: Establish goodwill.
Paragraph 2: Present the negative message, with the reasons.
Paragraph 3: Explain positive aspects and reestablish goodwill.

Nobody likes to receive bad news, so begin by establishing common ground or goodwill. Then deliver the negative message, giving the reasons if possible. (Research shows that people prefer to know the reasons behind the bad news.) To close, explain any positive aspects and reestablish the goodwill of your reader.

For Example

Let's take an example with which most of us have at least a little experience: employment rejection letters. (Empathy can be a very effective teacher.)

When we receive one of these bad news letters, we feel less disappointed when the employer gives reasons for not selecting us. At the other extreme is the employer who doesn't send any letter at all or sends only a form letter.

I once had an interview for a college teaching position. The interview started at 7:00 A.M. and ended eight hours later—almost. At that time, exhausted from the question-and-answer marathon, I had to give an hour-long lecture. Then the college held a wine and cheese party in my honor. Several people pulled me aside and said that I was the best contender for the position and that they would look forward to working with me. I left very tired, but satisfied that I had the job.

I waited for the call. After several weeks, I received a standard rejection letter in the mail with no indication of any reasons.

I was furious. I called the contact at the college and asked why I hadn't been selected. She said that they'd chosen a candidate with more than ten years' teaching experience; I had only three years' experience. If the college had simply indicated that reason in the letter, I wouldn't have felt so rejected.

It's easy enough to soften bad news by giving reasons—and by taking a positive approach. Let's look at a rejection letter that uses this approach (Figure 8-1, page 112).

We cannot avoid bad news—and we should never avoid communicating bad news when necessary. But we can make it easier on everybody involved if we accentuate the positive and provide reasons for the news.

Reacting to a Predicament

Sometimes tough situations mean dealing with people who are negatively disposed to you and/or your message, perhaps even hostile. They may be what are called "captive readers":

Weaver Insurance
1234 Michigan Avenue
Kalamazoo MI 49001
(616) 555-0190

May 19, 1998

Drew Lovell
321 Academy Street
Kalamazoo MI 49002

Dear Drew:

Thank you for interviewing with Weaver Insurance. We enjoyed meeting with you and discussing career opportunities.

After reviewing the results of your testing, we concluded that the "fit" here with the rest of our team wasn't perfect. Your results indicate that you are a self-starter and highly analytical. We were looking for a slightly different mix.

We hire for several public relations positions each year, and generally these employees move quickly to management levels. We'd like to consider you for the next available position, because each department requires a different team composition.

Thank you again for the time you spent with us. We look forward to speaking with you again in the future.

Sincerely,

Grant Crawford

Grant Crawford
Human Resources Manager

Figure 8-1. Rejection letter

they must read your report. You're encountering captive readers, for example, when you send a memo to your employees or to a vendor who needs your business. Your readers could also be hostile because they disagree with your position on an issue or because they dislike your company or even the entire industry. Finally, the hostility could be because the recipients of your message just don't like or trust you personally.

Whatever their reasons for being negatively disposed to you and/or your message, you face a difficult challenge when you're dealing with both a tough situation and tough readers. So what can you do to reach that hostile reader?

The best approach may be the following five-step strategy:

1. Announce your position and briefly state your reasons.
2. Recognize the reader's possible objections to your position. Agree on some point if you can. Use your research to recognize the reader's position.
3. Refute each objection by presenting your evidence, especially if you have statistics.
4. Argue your position, developing each point.
5. Conclude with a memorable statement, repeating any specific actions you want the reader to take.

Note that in step 1 you state your reasons briefly. That shows you to be reasonable, but not aggressive. Step 2, which is very important, lessens the adversarial nature of the relationship between you and the reader, because you show that you understand the objections to your position—and even agree to some extent. Providing reasons for agreeing is an indication of your sincerity and good faith: you're not just playing a game, trying to appease the reader.

With step 3, you begin asserting your position by arguing against your reader's position. We emphasize presenting reasons and evidence, particularly hard facts such as statistics. Although step 4 is to argue your position, you should maintain the same tone as in the first part of your letter or memo—reasonable, pleasant, and courteous. The force of your argument should come through your reasons, not your tone. Finally, step 5 closes the discussion decisively, with a statement that moves beyond words to actions. Figure 8-2 (page 114) is an example of a letter to a hostile reader.

When you're writing to readers who are negatively disposed to you and/or your message, you may want to hit with force, to overcome the opposition. However, an effective manager recognizes that the real challenge is not to win but to improve the relationship. And that takes great writing.

Above All, Do No Harm

The Hippocratic Oath that guides medical doctors advises them that their first duty is to do no harm. Although managers

Metro Magazine
2718 Haight Avenue
Cherry Hill, NJ 08034
(201) 555-0111

April 12, 1998

Jon Williams, President
Easttowne Automobiles
123 Lancaster Boulevard
Cherry Hill, NJ 08031

Dear Mr. Williams:

I stand behind Billy Joe Greene, editor of Metro Magazine, in his decision to run the article on the rising cost of automobile repairs in the April issue. I support him because our magazine exists to serve our readers, who are both consumers and businesspeople in our community.

I understand why you questioned that decision in the letter you sent to Mr. Greene on April 2. You feel that Metro Magazine should promote local businesses—or at least not question their practices. You also believe that, as a longtime advertiser in our magazine, you have earned the right to review and approve or reject any mention of your dealership in articles we plan to publish.

I appreciate your concern about the purpose of our magazine, and I agree with you that we should promote local businesses. In fact, we feel that we've been very supportive of the business community. You may recall in particular our feature piece in October 1997 on new car dealerships, in which the reporter focused on your Easttowne Automobiles. We certainly are very grateful for your advertising business and we'd like you to feel enthusiastic about every article we print. In fact, I checked with our sales representatives just this morning and they estimated that some 75% of our advertising partners have expressed interest in reading at least some of our articles before they go to the printer.

Unfortunately, however, we just can't allow even our best partners to review and approve or reject articles. Our production schedule is simply too tight to allow us to check with our advertisers on every article. In fact, the magazine copy has gone out late to the printer three times in the last six months. We also have a policy that prohibits prior review, as the journalists call it, because we want to protect our writers against any kind of outside influence. We owe that objectivity to all our readers.

I'm very glad that you took the time to write to the editor of Metro Magazine. I invite you to write directly to me if you have any further concerns or questions. We can work together to make our community business magazine better for all of us.

Sincerely,

Virginia Appleby

Virginia Appleby
Publisher, Metro Magazine

Figure 8-2. Letter to hostile reader

don't take any oath of office, that advice applies equally well to you and how you communicate, especially in tough situations: above all, do not offend.

Imagine receiving a past-due bill that states, "If you do not remit this payment by May 1, we will send this information to a collection house." How do you react to this method of delivering negative news? Probably not too well.

Unfortunately, negative news is often delivered harshly. But people find that approach offensive: no one likes to be intimidated, insulted, or threatened. In fact, that approach may not even work. Sometimes readers may become so annoyed by such a past-due notice that they withhold payment to punish the company. They're at least likely to tell their friends and neighbors, which can't be good for business.

You can't avoid tough situations, but you can certainly avoid offending people with the manner in which you deliver bad news. Let's look at several examples of sentences that express something negative. Notice the difference that a slight revision can make.

If we don't receive your payment by May 1, you'll be billed an additional fee.

Please send us your payment by May 1 to avoid the late fee.

Unless you send us this information, we will not process your claim.

We will process your claim as soon as we receive your information.

Until you receive our permission, you may not begin the job.

You may begin the job as soon as you receive our authorization.

When you need to convey a negative message, stop for a moment to think about the reader. A little empathy should help you find a more positive way to deal with a negative situation.

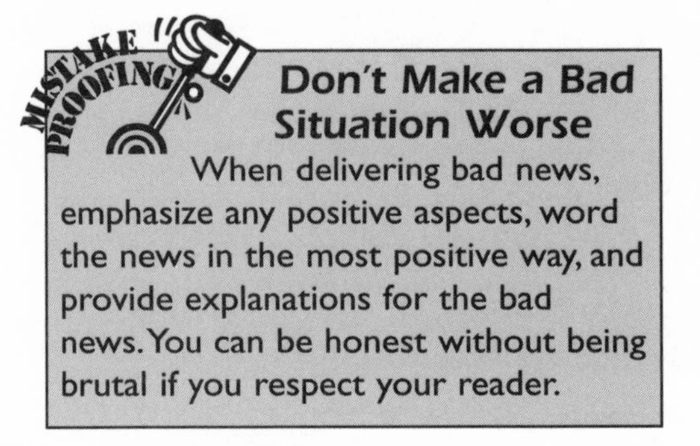

Don't Make a Bad Situation Worse

When delivering bad news, emphasize any positive aspects, word the news in the most positive way, and provide explanations for the bad news. You can be honest without being brutal if you respect your reader.

Turning Negatives into Positives

Managers who communicate well are like medieval alchemists, those optimists who worked hard to turn common elements into gold. They failed, of course, but you've got a chance to succeed at modern business alchemy. A manager who writes well can often turn negative situations into positive advantages.

It's essential to be able to deliver bad news or resolve problems. It's important to be able to keep a bad situation from getting worse. But it's great to be able to do all that—and improve relationships, build business, and actually come out ahead.

How can you achieve that sort of reversal of fortune? By emphasizing relationships.

This emphasis is the key to the success of the five-step strategy outlined above. When you start off by giving reasons for your position in step 1 and recognizing the validity of the reader's possible objections in step 2, you're treating that person with respect. As noted earlier, you're trying to rise above any adversarial relationship that separates the two of you by building on common ground. You start with any basis of agreement—if only the implicit wish to keep any disagreement civil and rational, founded on facts and logic. Your tone should be reasonable, pleasant, and courteous.

People can often overcome the differences that separate them—if they feel that they're alike in the way they deal with those differences. Before you begin writing a difficult letter or memo, especially if the recipient may be negatively disposed, think about how you would like your relationship to be. If you aim at building that relationship—whether it's with your employees, other managers, your customers, business partners, or the media—and not just at resolving the immediate problem, you're more likely to be able to turn negative situations into positive advantages.

Competence and Confidence

As I stated in opening this chapter, few managers feel comfortable dealing with tough situations. But maybe the suggestions I've made here will help you become more competent in handling such situations and feel more confident in doing so.

Consider the scenarios outlined at the start of the chapter. Think about the memos, letters, and reports you'd write to deal with those problem situations.

- *How would you rid your department of the lax employee—and perhaps notify her coworkers of her departure?*
- *How would you respond to the letters from those unhappy distributors—and notify the president of your company, who may be getting word from other sources about the unsatisfactory shipments?*
- *How would you break the bad news to the three employees you have to lay off, knowing that the mood in your department is already critical?*
- *What about that media mess? To control the damage, what would you write and to whom?*

Tough situations. But I hope you'd be more able to handle them well because of this chapter.

Angry or Frustrated Customers

No matter how upset a customer may get, never become emotional. No matter what a customer may write about your organization, don't get defensive. Emotions and defensive reactions only make a problem worse.

All of us have been angry or frustrated customers at some time. Use that experience to empathize with the complaining customer.

Manager's Checklist for Chapter 8

- ❑ When breaking bad news, phrase it in positive terms and give your reasons.
- ❑ A good strategy for delivering bad news is to start by establishing goodwill, then present the negative message and the reasons, and conclude by explaining the positive aspects and reestablishing goodwill.
- ❑ Write more effectively to people who are negatively disposed to you and/or your message by being reasonable, respectful, and sensitive to their positions and their reasons.
- ❑ Avoid offending—and especially avoid words that intimidate, insult, or threaten. Treat your reader as you would like to be treated.
- ❑ Turn negative situations into positive advantages by focusing on building relationships, not just on resolving problems.

Persuading Your Readers

Most people are hard to persuade and somewhat inert; even if you persuade them, you must get them to act. Usually when we seek to persuade people through our writing, we hope to provide information, change attitudes, or even change behavior. As I discussed in Chapter 4, the readers may have a negative attitude about your message or even about you. You must overcome that predisposition to gain their attention.

Obviously, persuading readers to think, feel, or behave differently is a greater challenge than simply conveying information. Don't feel discouraged. Persuasion is a difficult science.

For example, a student at Temple University gave a powerfully written presentation about ten years ago on raising the speed limit from 55 mph to 65 mph. At the conclusion, he asked for a show of hands to indicate how many wanted the speed limit raised. The entire class of thirty students responded affirmatively. Then he asked how many would sign a prepared letter to the appropriate members of Congress urging them to change the law. Only ten people raised their hands. Then, when he asked how many would mail it, only three students volunteered to take action. So even though thirty people were in favor of the change, he could count on only three to act.

How to Persuade in Three Difficult Steps

The basic strategy of persuading can be expressed as three steps:

- Establish common ground.
- State the problems and solutions.
- End with a strong action close.

Of course, it's much easier in theory than in practice—especially when you're trying to use words on a page or a screen without being able to "work the room."

But you can improve your chances of success by understanding what those steps involve. A more sophisticated way of looking at the process of writing to persuade is as follows:

- Gain the attention of your readers by presenting a benefit—or at least by establishing goals of mutual interest or common ground.
- Define the problem that will be solved if you succeed in persuading the readers.
- Explain the solutions, showing how the advantages of the solutions outweigh any negatives.
- Enumerate the benefits for the readers.
- State the specific action you want the readers to take.

Let's try an example of persuading readers. Suppose the president of your company has promised to help a local environmental group organize a drive to clean up an area river. They need volunteers and money, so she's asked you to write to the presidents of the 100 largest businesses in the community to enlist their support.

You know that your letter will go out to some people who may not be open to your invitation to support the cleanup. Some probably won't feel very concerned about the river. Others might care, but won't believe it's an appropriate cause for their companies. Still others may be reluctant to participate because the environmental group has gotten involved in some political campaigns. How can you establish common ground or provide a benefit to gain the attention of people who don't share the interest of your company president in this cause?

You might start with some statistics on the recreational use of this river, particularly by showing the economic impact of this use. Connect your information to the readers by citing comments from local business and community leaders that they're likely to respect. Establish common ground, pointing out why they might want to listen to you.

Then go on to describe the problem of the polluted river and who's affected by it—fishers, boaters, and swimmers. Put faces on those people for your readers: those people are their neighbors, their friends, and maybe even their families and the families of their employees.

Describe how the company presidents will help solve the problem by encouraging volunteers or donating money for the cleanup. Dispel any constraints, any reasons why they might not want to support the effort. List the benefits for them, intangible as well as tangible. You might want to mention, for example, that the local media will be covering the event. Invite your readers to join in the effort by pledging volunteers and/or money.

This last step is crucial. If you don't specify the action you want them to take, you may persuade them but not achieve the results.

Ten Rules of Persuasion

According to *Power-Packed Writing That Works* (published by *communication briefings*, 1989), ten basic rules can help you persuade readers:

1. Know your readers.
2. Know what you can accomplish.
3. Anticipate objections.
4. Stress rewards.
5. Be familiar.

6. Be clear.
7. Ask for what you want.
8. Control the tone.
9. Clinch your argument.
10. Give them something to remember.

Let's look at each of these rules a little more closely.

Rule 1. Know Your Readers

The effectiveness of persuasive messages depends on the readers. Determine to what extent your readers seem active or passive. Active readers seek information, and may want to read your message. Passive readers need motivation and a hook to interest them.

If you know or suspect that the recipients of your message are active, stress features. Tie into their agenda. They might be positively inclined to your idea, so don't introduce many new thoughts. Try to picture the payoff—describe the results of your proposal.

If you believe that the recipients of your message are passive, discuss benefits. Give them a problem—particularly one that matters to them. Remember: adults learn best from examples. Quote others and provide examples from individuals that your readers respect. Use memorable phrases like slogans to reinforce your message. If you can identify your readers' agendas, priorities, interests, and values, you can appeal to what's important to them and be more persuasive.

Rule 2. Know What You Can Accomplish

Usually we want readers to think, to feel, or preferably to act. But don't expect a 180-degree change from one letter or other communication. If you can move readers from extremely negative to only somewhat negative, that's success.

If you want to change a position or an attitude, you must know its reasons, its roots. You need to understand the thoughts, feelings, beliefs, and whatever else is behind that position or attitude. Appeal to your readers by presenting information in such a way that it makes a difference to them.

Make your wording memorable.

Get to the underlying thoughts, feelings, and beliefs. Tap the power of emotions such as ambition, pride, guilt, greed, fear, a love of challenges, or a sense of community.

Changing behavior is the most difficult goal of persuasion. Know what's realistic and recognize that change may take a lot of effort. You need to understand the obstacles that make it difficult for your readers to yield to your persuasion.

Rule 3. Anticipate Objections

Many people have significant constraints, reasons they can't yield to your persuasive words and act as you'd like. We're all familiar with the many constraints in business—budgets, mission statements, facilities, federal and state regulations, limited vision, local ordinances, and so on.

Sometimes people won't react to your persuasive words because they don't recognize the problem or the opportunity. If you identify it for them and then offer ways to resolve the problem or take advantage of the opportunity, you may improve your chances of persuading them.

According to James E. Grunig and Todd Hunt, authors of *Managing Public Relations and Public Relations Techniques*, if you motivate people to take action, they'll find a way to do it. That is, they'll overcome their constraints.

But persuasion depends on the situation, of course. Here are four types of readers you may encounter:

- Problem-facing readers see the problem or opportunity and face few constraints. You can persuade these readers.
- Constrained readers recognize the problem or opportunity but face significant constraints. You'll have difficulty persuading these readers to act.
- Routinely behaving readers face few, if any, constraints, but they have trouble seeing the problem or opportunity. You must encourage these readers to take action.
- Fatalistic readers cannot recognize the problem or opportunity and are highly constrained. Trying to persuade these readers may be a waste of time and money.

If you know your readers, you should be able to anticipate their objections to your message and to identify their constraints. Then you can dispel their objections and minimize or eliminate their constraints one by one so they are free to be persuaded.

Rule 4. Stress Rewards

When you establish common ground or write about benefits, picture the advantages for the readers. Make the readers feel how they will be better off as a result of your proposal.

For instance, when trying to persuade management to set up a corporate intranet, what benefits could you cite for the company? Although employee morale might make your top ten list of reasons, management might have difficulty picturing that benefit. However, if you talk about increases in productivity because of the opportunity to work as teams on the intranet, management might picture higher profits.

As I noted earlier, if the recipients of your message are active, you can persuade them by discussing features. But with recipients who are passive or not predisposed toward your message, you must stress benefits to obtain and keep their attention.

Rule 5. Be Familiar

Plug into what your readers know, appreciate, like, and respect. In other words, press their hot buttons.

If you're citing facts, use a source that your readers consider credible. If you want to include quotes, get them from people that your readers respect. The more familiar the person you use, the greater your opportunity to persuade readers.

Here again, it's vital to know your readers. Build on what you know, not on assumptions. Beware of generalizations: all CEOs are not alike; all small-business owners are not alike. Understand the differences among people and you can persuade them appropriately.

Rule 6. Be Clear

If the people reading your words have to work hard to understand them and particularly to understand the benefits, your message won't be as effective and you may not move your readers to act. Make it easy for them to understand what you're saying.

Rule 7. Ask for What You Want

State what you want the readers to do. Use powerful verbs. Keep your sentences short. You might want to emphasize the action by using boldface or italics, by setting it off in a separate paragraph, or by repeating it.

If you know your readers, you should know how to make your point with them. Dare to go for what you want. That's your purpose in writing to persuade. Don't be subtle. Don't be shy. Make it easy for your readers to understand what you want them to do. Tell them, for example, "Sign the petition to increase the speed limit" or "Give your family a future with you."

If you've persuaded them to read this far into your letter, memo, or proposal, the last thing you want to do now is confuse them. Tell them what you'd like them to do.

Rule 8. Control the Tone

If you want to persuade the audience, use the tone that works best for them and that best suits the purpose of a given part of your message. What I mean here by "tone" is the concept of person.

The first person ("I") signifies authority. This tone might work if you are an expert or otherwise very credible to your readers: "I believe that this initiative would be best" or "In my opin-

> CAUTION!
>
> **Spell It Out**
>
> Don't assume that readers who are sympathetic to you or to your message will take action. Outline your reasons clearly and from their perspective, then clinch your argument and specify what you expect from your readers.

ion..." Otherwise, it could annoy the audience because they don't believe you or identify with you. You might confine your use of the first person to quotes from credible people and to a postscript, where you can add your personal, persuasive touch.

The second person (you) signifies familiarity. You already know that you can be more persuasive if you address your readers directly, as if conversing with them. When you use "you" rather than "I" or "we," you tend to write more from the perspective of your readers, to focus on their interests and needs, to emphasize the benefits for them of taking the action you recommend.

Use the third person (it, the company, he, she) when you want the tone to feel more objective. Sometimes objectivity works well in persuasion, particularly when you're presenting facts or supporting materials and you don't want to appear personally involved. However, avoid using the third person throughout your message unless it's a formal proposal, because the third person doesn't move readers as effectively as the second person or even the first person.

Rule 9. Clinch Your Argument

Stephen Toulmin, who developed the Toulmin Model and wrote *The Uses of Argument*, suggests that you can most effectively convince people of the value of your argument by using a three-step approach:

1. State your purpose.
2. Support your argument with evidence.
3. Close with a clincher—a reason for the argument.

If your readers share your reason, your argument will be stronger, of course. Use common beliefs or attitudes as clinchers—who would disagree, for example, with such values as consistency, quality, productivity, or profitability? If your readers agree with your clincher, they'll probably agree with your argument. Signal your conclusion with markers such as "therefore" or "thus."

Let's look at a very simple example in the box just below of a three-step argument excerpted from an internal memo.

Purpose: We need to expand the market for personal computers to include people who cannot afford to buy machines that cost more than $1000.

Evidence: We've saturated our current market for computers priced over $1000 that consumers will not need to replace for at least three years. Yet millions of people in this country do not own a personal computer now nor do they plan to buy one in the near future.

Clincher: Considering this evidence, therefore, you'll agree that the growth and profitability of our company depends on what we do now to develop and promote computers that can be sold at prices ranging between $500 and $800.

Rule 10. Give Them Something to Remember

Sum up your message in a brief, memorable phrase. Research indicates that you must get the readers to remember your message if you want to change their opinions and affect their behavior. Focus on a key point in your argument, for example. (Most people will long remember the clinching conclusion in a famous court case: "If it doesn't fit, you must acquit.") Or focus on a significant benefit for your readers. (Consider the persuasive power of such old political slogans as "A chicken in every pot.")

Television has made this rule second nature to most people in power, because this rule is the essence of the ubiquitous sound bites that can influence public opinion on almost any subject imaginable. It's human nature to want it all "in a nutshell," so close by giving your readers something to remember.

Use a Problem/Solution Format

As I previously mentioned, many people like to solve problems. If you present your persuasive message using a prob-

Persuasive Punch

Smart Managing

To give readers something to remember, use words that pack a persuasive punch.

According to a Yale University study, the most persuasive words are: discovery, easy, guarantee, health, love, money, new, proven, results, safety, save, and you!

lem/solution format, you may be able to hook your readers.

People usually like choices. If you give them a problem, offer several solutions. In the case of your river cleanup event, you could suggest three actions in your letter to the company presidents:

1. Send a check to the environmental group.
2. Join volunteers from the community to remove debris from the river.
3. Encourage your employees to volunteer for a few hours.

If you'd like to push one of the solutions as preferable to the others, you may position it between two less appealing solutions. This strategy is especially effective if you put the least attractive option first. For example, if your river cleanup really needs money more than volunteers, you might stack your choices like this:

1. Show up ready to get down and dirty in hip waders and an old shirt.
2. Send a tax-deductible check to the environmental group.
3. Personally encourage your employees to give up part of their weekend to volunteer for a few hours.

You want to promote choice 2, so you set it between less convenient choices—which you word to make them seem even less appealing to the company presidents.

Choices make it easier for your readers to act. If they have only the basic choice between acting and ignoring your plea, it's easy to do nothing. But choices allow them to be persuaded, even if most of them opt for the easiest choice.

On the other hand, suppose you believed that the problem had a greater appeal for the recipients of your message. Maybe it presents more business benefits or maybe you have the endorsement of popular community leaders, such as the mayor or the publisher of the local business magazine. With that power behind your cause, you might make your message more effective by not presenting any choices. The simplicity serves to challenge the company presidents: they're going to choose between joining in or being left out.

Write Powerful Proposals

A proposal is a specific, highly persuasive document. The basic format for an effective proposal consists of the following essential parts:

- **Introduction.** Discuss the need, problem, or opportunity. State your goals. Outline the benefits for the people reading the proposal.
- **Body.** Present the actions you propose. Describe the options. Sketch out a timetable and a budget.
- **Conclusion.** Propose how the readers should meet the need, solve the problem, or take advantage of the opportunity. Emphasize how they will benefit from taking action.

> **Put Faces on Your Facts**
>
> TRICKS OF THE TRADE
>
> You should use facts to persuade your readers, to appeal to their minds. But never underestimate the power of appealing to their emotions as well. When you represent benefits in terms of dollars and percentages and other data, depict those benefits in terms of how they'll affect people.

How can you make your proposals more powerful? Try the following suggestions:

- Use a memorable title to attract attention, generate interest, and inspire action.
- Introduce the proposal with a message that highlights the need, problem, or opportunity. Support your statement with facts and/or quotes.
- Begin the body with a purpose to help readers understand your proposal and how the benefits serve their interests.

- Keep the proposal concise. Place complex supporting information in an appendix.

According to *How to Get Results with Business Writing*, (*communication briefings,* 1992), powerful proposals clearly state the purpose and the problem or opportunity, present a convincing and timely solution, and show that the benefits exceed the cost. An executive summary can also make a proposal more effective by giving readers an overview of the most important information in a short, easy-to-read format.

Let's look at an example of a proposal.

Scenario: Maxine Ott is the public relations director for SKE Chemical. She knows that the owners of a local public golf course want to sell. She thinks her company could benefit from owning the golf course. The proposal she writes for upper management to recommend buying this course follows:

TO: SKE Chemical Executives
FROM: Maxine Ott, Public Relations Director
RE: Proposed Community Relations Plan
DATE: May 14, 1998

[introduction]

[need/problem] Currently, we suffer negative public opinion from our surrounding communities because of the nature of our industry and because of people's fears about chemical disposal and waste in their own backyards. **[benefit]** We need positive community relations to continue to do business here. Consumers today care about the social conscience of companies and ruin corporate reputations by boycotting products or prodding the media to attack "big business."

Although we dispose of our chemical wastes safely, consumers—particularly in our community—fear a chemical accident. With potential picketers and the negative media coverage we've recently received concerning our proposed addition, **[goal]** we need a way to reach out to the community with a program or project that meets its needs.

[body]

[solutions/options] I've researched one idea that might prove

Continued on pages 131 and 132

financially beneficial for us at SKE Chemical while it offers the community a way of looking at us as a neighbor rather than a potential problem.

The owners of Hickory Run, a local public golf course, would like to sell their interest in the course and move on to other pursuits. They have enjoyed a profit from Hickory Run and now wish to sell the course so they can retire.

My research shows that 35% of community residents play at the course at least twice each year. In addition, 26% of our own employees enjoy golf and could become "members" if we bought the course.

My recommendation, after speaking with our CFO, John Dottermusch, is to purchase the course and make it a semiprivate course, offering memberships to our employees at a reduced rate. Perhaps more important, we could offer memberships to community members who desire the opportunity to make advanced tee times. The course would still offer its other tee times to regular public course players. I would develop an aggressive public relations/advertising plan to feature the change of ownership and community ramifications.

My research indicates that the community would respond favorably if we reached out to them and offered something they value.

[timetable] The owners would like to complete the sale by summer's end. We could enter negotiations in June and conclude by August.

[budget] Dottermusch believes that Hickory Run currently produces and could provide an appropriate return on investment for us. The prenegotiation purchase price is $21 million. Annual revenues currently exceed $2.5 million. Costs currently run $1 million yearly. Remember also that we would own sixty-four valuable acres in a growing community. And we would receive a charitable deduction for providing a community resource.

[conclusion] By purchasing Hickory Run, SKE Chemical would realize the following benefits:

- Improved community relations
- Minimized negative media coverage
- End to picketing
- Improved employee morale
- Financial gain
- Addition of real estate holdings

Continued on next page

- Charitable deduction
- Community goodwill
- Pride in SKE Chemical as a good corporate citizen

Please carefully consider this important proposal and provide your feedback to me by May 31. We need to act quickly before other investors begin negotiations.

This proposal uses a clear introduction, body, and conclusion. The intent of the public relations director in proposing the purchase is to build positive community relations by offering the community a service it values.

Notice that the introduction states the problem (negative community relations), a benefit to solving the problem (improving corporate reputation), and a goal (a positive relationship with the community).

The body presents a possible solution to the problem. The PR director supports her proposed solution with research results and the opinion of the CFO. She ends the body with a timetable and a budget.

The conclusion lists the many benefits of purchasing the golf course. The bulleted list format makes those benefits stand out for the readers. The PR director closes her proposal by specifying the action she would like, a review of the proposal by the end of May.

In this scenario, it isn't necessary to elaborate on the problem, because the executives are acutely aware of the company's image in the community. On the other hand, because it's not part of "business as usual" for the chemical company to invest in recreational property, it's essential to detail the specific benefits in the conclusion. Because she's simply trying to encourage negotiations, not necessarily to push the purchase, the director presents only the rough outlines of the financial picture. In fact, if she presented any more detail, she might distract the executives from the main thrust of her proposal.

To write the most effective proposal, you need to study the situation from the perspective of your readers. I cannot emphasize this enough. Study each aspect of your proposal separately. Where will you encounter problems with your readers? Where can you assume things will be easier?

If you believe, for example, that your readers don't know about the problem you want to solve or that they don't consider it to be very serious, then you need to put a lot of attention into explaining the problem and emphasizing its importance. But once they understand the problem, maybe you'll need only a sentence or two to present your solution.

Conclusion

I began this chapter by stating a problem: "Persuading readers to think, feel, or behave differently is a greater challenge than simply conveying information." That was how I established a goal that I hope was of interest to you—being able to write more persuasively.

I put most of my effort into explaining the solutions to that problem—and I hope I've shown that the advantages of the solutions outweigh any negatives. (Sure, it's a lot of work to write well, but it pays off when you can persuade people.)

I may not have done much to enumerate the benefits for you. But the very fact that you bought this book and you're now reading these words indicates that you're already aware of at least the basic benefits—aware enough that you're willing to work at improving your writing. (It's not that I know my readers, but that I know about self-selection: if you're reading this, it's because you're interested.)

Now, if I want to conclude this piece of informative and persuasive writing properly, I need to state the specific action I want you to take. First, review the manager's checklist. Second, try a simple exercise during those little moments throughout the day, like during lunch or on the commute. Think about something that you believe or like or want. Then pick a person you know, at random, and try to view your subject from that person's perspective. How could you persuade him or her to believe it or like it or want it? As this exercise becomes more natural, you'll be developing your ability to persuade with your writing.

Manager's Checklist for Chapter 9

- ❑ Remember that persuasion is difficult. Don't expect too much from one document—or try to pack too much into the package.
- ❑ Know your readers and know the specific context.
- ❑ Establish common ground or state benefits to gain the attention of your readers.
- ❑ Don't just provide information or aim at changing attitudes. The most effective persuasive writing produces action.
- ❑ Always state the specific action you want the audience to take.
- ❑ Anticipate questions, concerns, constraints, and objections.
- ❑ Stress benefits for your readers.
- ❑ Conclude with power. Tell them what you expect from them and give them something to remember.

Reports, Reports, Reports

Reports are the most common vehicles for writing to provide information and often persuade the readers. You may write many kinds of reports, but they can be categorized into the following four types:

- ***Occasional report:*** to alert or update on a situation.
- ***Activity report:*** to sum up a trip, a conference, a meeting, or any other event.
- ***Status or progress report:*** to give a general review of activities in a department or progress on a particular project.
- ***Formal report:*** to provide a comprehensive overview.

These types of reports differ in format and complexity. We'll consider the differences among the reports after we discuss what they have in common and how you prepare for them.

Basics for All Reports

Whether you're sending off an occasional report to a colleague down the hall or preparing a formal report for the board of directors, a report is a report. If you think in terms of objectives, readers, and scope, you've got the foundation for a great report.

Objective

The objective is your purpose for writing the report—and it's not just because somebody asked you to do a report or because it's standard procedure in certain situations. As I stressed in Chapter 2, you need to establish your purpose when you write anything before you even touch a pen or a keyboard.

If you're writing a report, your general purpose is to inform and maybe to persuade. But that's only a start. You need to refine that general purpose into an objective, particularly one that will compel you to write more effectively and compel your readers to take a greater interest in your report.

Suppose you're writing an activity report. You might start by thinking, "I'm writing to inform others about the conference I attended last week." But that objective expresses only the what, without any sense of the why.

Push your thinking to include reasons, to generate interest. You might then come up with something like "I'm writing to inform others about the conference I attended last week because I learned about a trend that could be profitable for us if we accelerate R & D on that new product idea." Now you've got an objective that should inspire you to write your report with greater commitment and passion—and that feeling should generate greater interest among your readers.

That's how you move from your general purpose to develop your objective, a compelling reason to communicate through your report, to connect with your readers. The next step is to identify those readers.

Readers

As I emphasized in Chapter 1, whenever you write something, you have to identify the people who'll read it. Effective communication means connecting with your readers, with their knowledge, experience, and interests. That's just common sense, it would seem, and yet many managers write as if they were sending words off into the air with no particular target—"to whom it may concern." That's why their readers often don't seem too concerned about what was written.

Identify your readers and focus on what they know about the context for your report. Then you can attract their interest, meet their expectations, and deliver your information most effectively.

You've got to start with your objective—why you're writing the report. But if you don't identify your readers and think about why they should read your report, you'll have more trouble connecting with them and achieving the appropriate results.

Scope of Coverage

Once you've developed your objective and identified your readers and what they know about the context for your report, it's time to determine your scope. In other words, how much information should you give?

Scope is a question of balance—and generally of compromise as well. You've no doubt read reports that went far beyond what you wanted or needed to know, so you lost interest as the words went on and on. You've also probably read reports that left you hanging, wanting to know more, needing information that just wasn't there.

That's what happens when managers don't keep in mind and in balance the two parts of the information equation—what they want to give and what their readers want to receive.

It's easier, of course, when you're making a presentation orally. You pay attention to the faces and you either cut short your report or elaborate, depending on the feedback from their expressions. But when you present information in writing, you have to think in advance about the balance you should achieve.

Maybe you're enthusiastic and full of informa-

> **Information Equation**
>
> MISTAKE PROOFING
>
> The information equation is the relationship between what information you're sending and what information the intended recipient is ready, willing, and able to receive.
>
> You can greatly influence this equation by drawing and building interest in your report.

tion, but you suspect that your readers want less rather than more. So you need to compromise. The reverse could also happen; occasionally your readers want more detail or greater analysis in a report than you care to provide. Again, compromise.

If you're writing for more than one reader, especially if your readers have different expectations, needs, interests, time, and attention spans, you need to do even more planning. That's what we'll discuss next, within the context of the four basic types of reports.

Types of Reports

As I noted at the start of this chapter, business reports fall into four basic categories. Let's examine the differences among them.

Occasional Report

The occasional report is so much a natural part of working with others that you might not even think of it as a report. When something happens or you get some information, you may instinctively want to let others know. That's the occasional report, the "FYI" or "by the way" type of informative communication.

The occasional report uses simple and informal writing. It usually takes the form of a memo or an e-mail, the written media closest to conversation, and is often intended for a single reader or a small group.

Make your occasional reports short and direct. You can make them more effective by structuring them in the following format:

- Subject
- Introduction
- Information
- Reaction
- Close

Subject: A few words that serve as a headline to catch the reader's attention. It answers the reader's question, "What am I reading?"
Introduction: A brief paragraph, perhaps just a single sentence, that gives your reason for sending the message—from the reader's perspective. It answers the reader's question, "Why should I read this?"
Information: The essential facts, specifics, significant details—an objective presentation. It answers the reader's question, "What do I need to know?"
Reaction: Any feelings or opinions you might have, any questions that arise, any interpretations or speculations you might share, any suggestions or recommendations you might offer. You may not express any reaction: none may be necessary—or perhaps it may not be wise to do so. It answers the reader's questions, "So what?" and/or "OK, now what?"
Close: Summarize the essence of your report, emphasizing the most important facts and/or the main thrust of your reaction. Because many people tend to read the subject first, then the close, and maybe nothing else, use the close to draw your reader into the body of your message. It answers the reader's question, "Why should I spend any time reading this?"

Although managers may be tempted to dash off an occasional report, particularly with the speed and ease of electronic mail, such reports will work more effectively if you outline first. You can simply jot down a few notes for how you want to cover the five basic parts. A little thought might add a few minutes, but the results will justify the preparation.

Example of an Occasional Report

Figure 10-1 on page 140 is an example of an occasional report.

Notice how the sales manager uses the subject line to give the marketing director a compelling reason to read the report. Then he explains in the first paragraph his reason for writing this memo: he hopes to help meet a marketing need.

To: Suzanne Farris, Marketing Director
From: R.J. Martin, Sales Manager
Date: July 8, 1998
Re: Possible marketing vehicle for healthcare delivery system

You recently expressed some concern about how you could expand your promotional efforts to move the healthcare delivery system into new markets in 1999. I have a possibility for you.

While attending a regional sales meeting last week, I met Alan Franklin, a media consultant from St. Louis. He told me that Atwater Publications is launching a monthly newsletter for entrepreneurs this fall titled *What's New?* It will target owners of small businesses and start-ups. The newsletter will carry space ads. Alan told me that the ad manager is Monique Evans at 314/555-0101.

I wonder if you could use ads in *What's New?* to reach a peripheral market for our system.

I'm guessing that space ad rates might be low for the first few issues until the newsletter gets off the ground, so I wanted to let you know about this opportunity immediately.

Figure 10-1. An example of an occasional report

He presents the information as concisely as possible, with the basics and a contact. He follows up with his perspective on this information, that the newsletter could promote the product in a new market. (Because the marketing director will no doubt consider that possibility by the middle of the second paragraph, it's really not necessary for the sales manager to spell it out in the third paragraph—except to soften his suggestion diplomatically.)

He closes with subtle urgency, giving the reason why he's passing along the information now—and why the director should act quickly. Again, he uses sensitivity, so as not to give the impression of telling the director how to do her job.

In this short memo, the sales manager covers the five essentials:

- He headlines the *subject* to interest the recipient.
- He connects with the recipient in the *introduction*.

- He conveys his *information* concisely.
- He expresses his *reaction* to the information by making a suggestion.
- He *closes* with a subtle reminder of the importance of acting quickly.

Activity Report

You write an activity report to sum up a trip, a conference, a meeting, or any other special activity, to share any information and impressions—and to justify the expenses. As I mentioned above, you should focus not just on what happened, but also on why it matters to your readers.

The activity report should include the following elements:

- Event, place, and date
- Purpose
- Contacts
- Conclusions

Event, place, and date. Identify the trip, conference, or meeting. This information would seem only logical, but some managers are so focused on the event that they forget that others in the company may not be familiar with it. Also, spell out the names of any organization(s) involved: an acronym like "ABA-MAPA Conference" may mean nothing to your readers.

Purpose of the trip, conference, or meeting. Why did you participate in this event? If it's the most important national event for your specific industry, give that as your primary reason. But tie it into more specific needs; connect it with particular projects or other objectives. (If you participated only because of the importance of the event, you probably didn't prepare enough to get the most out of your presence there.)

Contacts. Whom did you meet during the trip, conference, or meeting? Identify them in terms of position and organization. Don't assume that the people reading your report will immediately recognize the names "A. B. Martin" and "Maryann Gibson" as impressive people. Also, specify the topics of your conversations in connection with your specific needs and interests if possible. (When you prepare for any special event, you should always establish a personal agenda that includes a list of the people you intend to contact and the topics you'd like to cover.)

If the event was a conference, also include the presentations you attended and what you learned from them. If the event was a meeting, outline the agenda, then focus on whatever was of significant interest.

Conclusions. What conclusions have you reached as a result of the event? You should have outlined this section of your report in your motel room immediately after the event or on your way home, while all the details were fresh. The purpose of this section of your report is to tell your readers how and why you and the company benefited from your presence at this event.

The activity report requires more planning than you might think. Be specific and concise in stating your purposes for the activity. Those purposes should also come through in the way you report on each of your contacts—and in the way you order them. Just presenting them in chronological order may be simpler for you, but it's far less effective. Help your readers understand the relative importance of your contacts.

The conclusion may require a lot of thought and planning. You need to synthesize what you learned from your contacts and present it in terms of your purposes for participating in the activity. If you didn't learn much, you should admit that and provide an explanation. Maybe you can offer some recommendations for better preparation next time, for example, or simply suggest that no one attend the conference next year.

An activity report allows you to show how you can think. Your outline structure will be simple, but you should put a lot of thought into what you include, how, and especially why.

Example of an Activity Report

Figure 10-2 (pages 143-144) shows excerpts from an activity report.

To: Marie Atkins, Wayne Jackson, Artie Munz, and Heidi Schwartz
From: Rosa Santoya
Date: August 20, 1998
Subject: Southeastern Conference for Magazine Publishers, Atlanta, July 15–18, 1998

Mark Andrews and I attended this conference to find out how other magazine publishers are dealing with the growth of the World Wide Web. We wanted to address in particular the following two key issues that are becoming critical concerns to all of us at Hobby Magazine Publications:

1. How are magazine circulations being affected by the Web?
2. How are publishers using the Web to attract and hold magazine subscribers?

Effects of the Web on Circulation

Mark and I attended a presentation by Anne Springer, vice president of the Association of Magazine Publishers. She presented statistics compiled by the AMP that show a decline in circulation of 12% over the last two years for a survey sample of 27 national general-interest magazines. (Attached is a copy of her handout.)

I talked with representatives from *Collector's Corner* and *Ham Chat*, two monthly magazines for hobbyists. Both representatives expressed concern about circulation loss and mentioned that they're considering going biweekly. *Collector's Corner* is also boosting its use of full-color photos, while *Ham Chat* is working more closely with regional amateur radio organizations.

...

Potential Uses of the Web

Mark met over lunch with Emily Moore, a Web designer who's helping several magazine publishers develop Web sites. She couldn't disclose any details, but she offered to consult with us for a few hours at no charge if we pay her expenses. She gave Mark the URLs of four sites she's designed, so we can check out her work.

...

Conclusions

It seems that we have good reason to be concerned about our circulation losses; other magazines are suffering as well. We may want to study

Continued on next page

three possibilities: publish more often, increase our use of photos and graphics, and work with hobby organizations, perhaps including offering discounts to members. It would also seem a worthwhile investment to check out the four sites designed by Emily Moore, then perhaps arrange a visit with her to determine if we can attract subscribers through a Hobby Magazine Publications Web site.

Figure 10-2. An example of an activity report

Rosa Santoya identifies the conference and presents the general reason for attending and two particular concerns—numbered for emphasis. She then organizes the body of her report according to those two concerns.

In describing the contacts made at the conference, Santoya provides only the most pertinent information. She sums up the presentation, then attaches a handout in case her four colleagues want more statistics. She reports on each contact in a separate paragraph to make it easier for her readers to follow.

In her conclusion, Santoya confirms the need to act on the company's concern over circulations, and she offers her recommendations. Her final paragraph is short and powerful—although she could have made it a little more effective visually by putting the recommendations in a bulleted list.

This activity report shows how a manager can cover the essentials—event, purpose, contacts, and conclusions—with substance, style, and strength.

Status or Progress Report

Status reports or progress reports usually provide periodical updates on the activities of a department or the results for a particular project. These reports may have a greater importance if management uses them to adjust personnel assignments, schedules, and budget allocations.

Status or progress reports generally consist of the following elements:

- Background

- Activities and results
- Costs
- Schedule
- Conclusion

Background. Establish the context for your report. If you're reporting on the status of your department, you might include a summary of what happened in the previous period. If you're reporting on the progress of a project, you might outline what's happened from the beginning and remind your readers of the purpose(s) of the project so they can better understand the significance of your progress report.

Activities and results. This section serves as the guts of your report. What have you done? What are the outcomes?

If you're informing your readers about a project, you can easily answer these questions. You just tell them about everything that's moving you forward toward your objective. You'll also need to mention any problems—and what you've done to resolve them.

If you're giving a status report of your department, you'll want to provide updates on each of your product lines or service areas. But you can better keep the attention of your readers if you can mention unusual activities—maybe some idea that you're developing for a new product or service, maybe an award your department received.

Costs. Sum up whatever it took to get the results you outlined in the previous section. You should compare the actual costs with budgeted costs. For a project, you should also mention projected costs for the next phase or the remainder of the project, with comments on any changes in the initial budget projections.

Schedule. Engage your readers with a look into the future. This step is particularly important when you're reporting on a project. But the future is important even in a departmental status report—if only to show how a good manager always thinks ahead and plans for success.

Three Quick Suggestions

- Anticipate questions that readers might have.
- Use lists—numbered or bulleted—to emphasize key points or important facts.
- As you write, keep in mind this question: "What do my readers need to know?"

Conclusion. End strong but short. A single paragraph puts your report into perspective. If you feel optimistic, express that feeling with your reasons. If you're concerned about possible obstacles, show how you intend to overcome the obstacles. Avoid exaggerated promises, because your readers will probably remember what you put in your conclusion.

Your report will be much better and far more impressive if you first outline carefully what you want to include. If you just start writing, you can almost guarantee that the report will be too long as well as too boring or difficult to understand. Some managers treat status or progress reports as routine—until they find themselves fighting against budget cuts or worse.

Take these reports seriously. Plan them well and you can get excellent results—routinely.

Example of a Progress Report

Figure 10-3 on page 147 is an example of a progress report.

Pete Parker starts with a general overview of the project. (He uses the same opening paragraph in every report.) He then summarizes the activities and results, followed by a brief statement of major costs and a comparison with budget allocations. Next, he outlines the project schedule for the upcoming months. Finally, he concludes positively with an assurance that the Gas BBQ Deluxe will launch as expected.

Most progress or status reports do not require a lot of detail. However, when the engineers create the polymer needed for the BBQ, Parker may well add a paragraph to announce that breakthrough and perhaps mention other commercial possibilities for the new polymer. Also, when the marketers finalize the outlines of their promotional campaign, he might

Monthly Progress Report on Gas BBQ Deluxe

May 1998
Pete Parker, Director of Product Development

We have been developing the new line of high-end chain saws since February 1998, with completion set for October 1998. This project involves design engineering and market research to determine how we can produce and promote a high-end model to expand our current product line of the original Gas Saw and Gas Saw Lite, and the five specialized models.

In April our engineering team tested seven polymer plastics for various parts of the saw. The results were unsatisfactory, but we hope to create a blend of three polymers that will provide greater durability without increasing the weight or production costs. We also expect to make extensive use of recycled plastics, which should lower costs and help our marketing efforts. In April the marketers concluded a survey of consumer preferences for features, size and weight, colors, and price. We are currently analyzing the results and will present them in our report next month.

The costs of building and testing prototypes were $75,000, just above our budget projection of $72,000. But at the three-month mark we are close to the $210,000 projected. The cost of the survey was only $8,000, so our total marketing expenses were $18,000, well under the $23,000 budgeted.

In May we will be testing our polymer blend, continuing to improve the basic design, and analyzing our survey results. We then hope in June and July to design the new features indicated by the survey results and to develop our marketing strategies for the launch.

We expect to be on schedule then to begin production of the Gas Saw Deluxe by late October, allowing distribution in early November in anticipation of the holiday sales season.

Figure 10-3. An example of a progress report

add a paragraph to cover that development in a little detail.

But for the routine report, a single paragraph is enough for each of the elements—background, activities and results, costs, schedule, and conclusion. It's a good idea to provide a heading for each section, especially if you have more than a single paragraph in each.

Formal Report

The formal report is the usual vehicle for organizing and conveying important information that's needed by different people in an organization. The format typically consists of all or most of the following elements:

- Cover
- Title page
- Abstract or executive summary
- Preface
- Table of contents
- List of illustrations or figures
- Body
- Conclusion
- Recommendations
- Glossary
- Appendix
- Endnotes
- Sources or references

Structurally, the formal report is much like a book, only on a smaller scale. The organization and the peripheral matters allow people to more easily access information of possible interest to them without reading the entire report. If you're preparing a formal report, you should thoroughly outline your contents in advance, as I recommended in Chapter 2.

> **Smart Managing**
>
> **Make It Count!**
>
> The cover letter or letter of transmittal provides an opportunity for you to talk to the readers. You can mention relevant details contained in the report and even provide a brief summary. In writing the letter, be sure to appeal to the readers' needs and interests. Use your letter to draw them into your report. Make it count!

Cover. The first page is a cover letter, either a letter of transmittal (if the report is for external use) or a memo (for internal use). It may be an explanation of the circumstances for the report, telling why and how you prepared the report, or it may be simply a note to the recipients.

Title page. The title page usually states the name of the organization, the division or department, the title of the report, the date, and the name(s) of the author(s). Again, the information will vary according to the circumstances and whether the report is meant for internal or external use.

Abstract or executive summary. This short overview of the report outlines the areas covered and the main points and offers a summary of the conclusions. Unfortunately, many recipients of a formal report read only the abstract or executive summary and the conclusion, so put the essentials here—and try to arouse interest in reading the entire report.

> **Short but Strong**
>
> TRICKS OF THE TRADE
>
> When writing an executive summary or abstract, keep in mind a few key principles. Keep it short—generally about 250 words or less. Make it strong—those few paragraphs should contain the essence of your report and make sense as a separate document.

Preface. This brief statement explains the background for the report, the objective, and the scope. Because the people who receive your report may stop reading at this point, try to interest them here in reading further. Use the CPO approach discussed in Chapter 6 to provide a context, state your purpose, and show the organization. You may increase the draw by summing up at least some of your conclusions and recommendations here.

Table of contents. This listing should indicate with page numbers the report's main divisions, heads, and subheads. Depending on the length and complexity of the report, you may not need a table of contents.

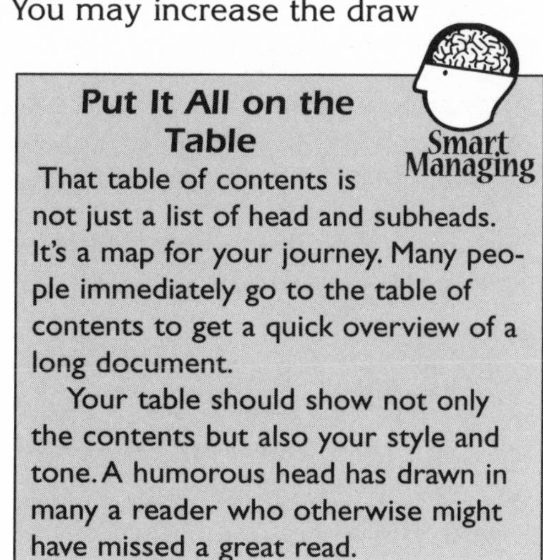

Put It All on the Table

Smart Managing

That table of contents is not just a list of head and subheads. It's a map for your journey. Many people immediately go to the table of contents to get a quick overview of a long document.

Your table should show not only the contents but also your style and tone. A humorous head has drawn in many a reader who otherwise might have missed a great read.

List of illustrations or figures. This list serves as a guide to the graphics; it provides quick and easy access to graphs, charts, and other figures. Of course, if the report contains few or no figures, this list is unnecessary.

Body. The body is the report proper, marked by heads and subheads, with the pages numbered. Refer back to Chapter 6 for more discussion about writing the body of a report.

Conclusion. This section provides a summary of the main points of the report and offers opinions and further considerations. You'll want to include your recommendations here, unless you want to emphasize them by putting them under a separate heading. Because many recipients of a formal report may read only the abstract or executive summary and the conclusions, write these two parts with the greatest care.

Recommendations. This section is unnecessary if you include your recommendations among your conclusions. However, if you want to give your recommendations extra punch, put them here, under a separate heading, perhaps in a bulleted list.

Glossary. If you use any terms that your readers might not understand easily, particularly technical terms, list them here with a definition. The glossary is particularly important if your report is for external use.

Appendix. This section contains statistics, tables, and any other information that might interest readers but doesn't fit appropriately in the body of your report.

Endnotes. Traditionally, notes to accompany your text are included here as endnotes. However, such information better serves your readers as footnotes at the bottom of the appropriate pages. Word processing programs have taken the effort out of using footnotes, so you can readily make your report easier to read. However, don't try to make footnotes long and detailed or to use too many of them. That style smacks of college term papers and theses.

Sources or references. The last section is a listing of any articles, books, reports, or other sources you used in your report. You might also include suggestions for further reading, especially if your report is for a wider group of readers within the organization or for external use.

Using Visuals

Don't forget to place each of your visuals near the text that describes it, making the visual accessible to the reader. Nothing irks readers more than having to shuffle ahead to the appendix while trying to read a description of the visual in the body of the report.

A formal report contains a lot of elements. That means that the points we discussed in Chapter 5 about using visuals, white space, and headings are even more important. The longer the report, the greater the need to think beyond the words.

Example of a Formal Report

Figures 10-4 through 10-6 show the cover, title page, and executive summary from an example of a progress report. I've omitted the body because we've already discussed how best to organize and write the text of a longer document in Chapter 6. I've also omitted other parts that are relatively simple and

TO: Alice Green, President
FROM: Brian Charles, Manager, Public Relations
DATE: August 11, 1998
SUBJECT: Public Relations Report

The attached report provides an account of our public relations efforts for the fiscal year 1997–1998.

Although Acme Chemicals, Inc., suffered some image problems because of the several minor accidents that occurred at our processing plant, we are pleased to note that the reputation of our company in the community is improving gradually.

I look forward to meeting with you next week to discuss any aspect of this report in greater detail.

Figure 10-4. Cover memo for a formal report

PUBLIC RELATIONS REPORT, 1997–1998

Acme Chemicals, Inc.
Office of Public Relations
Brian Charles, Manager
July 21, 1998

Figure 10-5. Title page of a formal report

PUBLIC RELATIONS REPORT

EXECUTIVE SUMMARY

This report provides an account of our public relations efforts for the fiscal year 1997–1998, including results from the promotional campaign that we began last year.

Because of the three minor accidents at our processing plant resulting in evacuation, we have conducted crisis management techniques, including community relations, media relations, and safety training for employees.

We have also realized the following results detailed in this report: increased community involvement and support, decreased negative media coverage, and image restoration resulting from new safety procedures ...

Figure 10-6. Executive summary of a formal report

familiar to all of us because they're included in many books—table of contents, list of figures, glossary, appendix, and references.

Managers often feel nervous about preparing formal reports. Such reports may be more complex because of all the elements to be included, but they are no different from any other informative business writing. Just keep in mind the essentials of organizing, focusing on your objectives, and writing for your readers, as we've discussed throughout this book, and you have no reason to be nervous about writing formal reports.

Evaluating Your Report

One last thing: after you've written your report, check it out from the perspective of your reader. Read it through as if you were receiving it. Do you find it interesting, informative, and easy to understand?

Occasional Report

Evaluating an occasional report is simple. Ask yourself the following questions:

- Do I focus on the subject as stated?
- Do I appeal to the reader's interests and/or address the reader's needs?
- Do I provide just the essential information?
- Do I make it easy for the reader to use this information?

You should also check over your grammar, spelling, and punctuation, of course—even if your report is "just going to a colleague." Always write everything as if it were going out to everybody in the company—because the written word is surprisingly durable and mobile.

Activity Report

If you're preparing an activity report, evaluating your writing is only slightly more complicated. Try the following questions:

- Did I sufficiently identify the activity?
- Did I give the general and particular reasons for this activity?
- Did I organize this report so that the readers can easily understand what happened, what I learned, and why this information is significant?
- Did I provide all the essential information for each contact or other source?

- Did I focus each part of the report on my reasons for participating in the activity?
- Did I attach additional documents, such as presentation handouts, background materials, and any other supporting information?
- Did I state conclusions and offer recommendations for using the information?

Once your report passes this test for focus, organization, and structure, check your grammar, spelling, and punctuation. Make sure your sentences are short and strong.

Connect with your readers and influence them. You're trying to inform them, yes, but also to show them the importance of participating in that particular activity. Remember: an activity report serves as both a cost-benefits analysis—how you'd rate the value of the activity—and as a performance appraisal—how you'd rate your use of this "information resource." Writing an activity report may be a routine procedure, but don't treat it casually.

Status or Progress Report

You should evaluate a status or progress report as you'd evaluate an activity report, with two differences.

- Provide a context for your status or progress report by referring to the past and to the future. Remind your readers of what your department or project team has done up to the current period and outline what you plan to do next.
- Accentuate the positive. I'm not recommending that you be a "spin doctor," but only that your status or progress report reassure your readers about your department or project. (If you can't do this honestly, what does that indicate about your ability to manage the department or the project?)

Formal Report

A formal report is very much like a status or progress report, so you evaluate it in essentially the same way. However, you need to perform two more steps to complete your evaluation.

First, check over your organization and format carefully:

- Make sure you included all the necessary elements, from cover to appendixes and references as appropriate.
- Compare your table of contents and list of illustrations against your heads, subheads, graphics, and page numbers. Verify all footnotes—are they necessary? Are they accurate? Are they complete?
- Inspect your figures by applying the criteria we discussed in Chapter 5.
- Read your preface, abstract or executive summary, and conclusion again, carefully and critically. Will someone reading only those three parts of your report acquire the essentials?

The second step is to use the Gunning-Mueller Fog Index test that I presented in Chapter 1. How readable is your report? Remember: it's not simply a matter of writing so that your readers understand your report, but so that they can understand it easily. Keep your sentences and paragraphs short and strong.

I also recommend asking a colleague to read your report. Encourage your test reader to be critical, to mark any areas in which you could improve the organization or the writing. Emphasize that the colleague not just find mistakes, but also suggest ways to make your writing more effective.

Reports: The Bottom Line

Whatever type of reports you write—occasional, activity, status/progress, or formal—your purpose is always to inform and often to persuade as well. That means you must connect with your readers, compel them to read and understand, and make your information come through as effectively as possible. The bottom line is always the extent to which your report fulfills your purposes.

Manager's Checklist for Chapter 10

- ❑ You'll use these four basic types of reports: occasional reports, activity reports, status or progress reports, and formal reports.
- ❑ Whatever type of report you write, think in terms of objectives, readers, and scope of coverage.
- ❑ Before you send or submit any report, read it from the perspective of the recipient(s), as if you were receiving it.
- ❑ Any report will be more effective if you focus on the specific subject, appeal to the reader's interests, address the reader's needs, provide the essential information, and make it easy for the reader to use that information.
- ❑ Always check over your style, grammar, spelling, and punctuation—no matter what type of report you write.

Postscript: Top Ten Tips for Writing Well in Business

The postscript of a letter or memo, like the conclusion of a longer report, is usually second in importance only to the beginning. I conclude this book with a postscript that is my attempt to present the basics for improving your business writing.

To synopsize the many principles and suggestions for managers in this book, I've devised the Top Ten Tips for Writing Well in Business. These tips should help you keep in mind the basics of writing well.

Top Ten Tips for Writing Well in Business

1. Know your readers.
2. Feature the "you attitude" and stress benefits for the readers.
3. Know your single communication objective or purpose.
4. Be clear, economical, and straightforward.
5. Use subject lines, indentation, short opening paragraphs, and postscripts.

6. Write strong introductions and conclusions.
7. Use headings, white space, and visuals in longer documents.
8. Write actively (subject-verb-object) rather than passively.
9. Avoid negative writing.
10. Use the power of persuasion to influence readers.

1. Know your readers.

To write effectively, you must know your readers. The more you know, the better you can focus your message on the individual or the group. Find out about such factors as age, gender, interests, values, attitudes, and knowledge of the subject.

If you don't know much about your readers, try to at least categorize them into one of four types:

- layperson—knows little about the subject
- expert—extremely knowledgeable about the subject
- executive—cares primarily about bottom-line information
- user—needs to understand in order to act

When in doubt, write so that the average 12-year-old could understand.

2. Feature the "you attitude" and stress benefits for the readers.

Motivate your readers from the start by writing from their perspective and emphasizing what's in it for them. A "me attitude" may condemn your masterpiece to the trash can. Whether you're writing an e-mail, a letter, or a report, focus on your readers.

3. Know your single communication objective or purpose.

Why are you writing? To inform your readers? To persuade your readers? To give instructions to your readers? To record an activity? It's more effective to focus on a single objective. Determine your purpose, then keep that purpose in mind as you write.

With every paragraph you write, ask yourself, "So what?" If that paragraph doesn't serve your purpose by taking you closer toward your objective, don't use it.

4. Be clear, economical, and straightforward.

Get your message through to your readers. Avoid wording that could distract or confuse.

Keep your sentences and paragraphs short. In general, sentences should average only 17 words or fewer. Paragraphs that run over seven lines or so tend to put off many readers. There's no magic formula, but the 17-7 guidelines should help you make your words work more effectively.

Structure your sentences to convey their meaning most directly. That's usually subject-verb-object with strong, active verbs.

5. Use subject lines, indentation, short opening paragraphs, and postscripts.

Attract your readers from the start and hold their attention until the end. Announce your focus with a subject line that grabs the readers. Indent the first line of your paragraphs to draw readers into your words. Make your first paragraph short to minimize the "initial investment." A postscript can make a big difference, especially when your purpose is to persuade, since many people read that last part of a memo or letter immediately after the subject line or first paragraph.

6. Write strong introductions and conclusions.

People often try to minimize their reading. They may want to read only the beginning and the end of a document—an executive summary or abstract, an introduction, and/or a conclusion.

Your introduction and conclusion must provide the essentials of your document and motivate the reader to at least skim the entire document. A strong summary makes a document more memorable, so it's likely to get better results. Repeat your main points to stick in the reader's mind.

7. Use headings, white space, and visuals in longer documents.

Guide your reader through your document with informative, interesting heads. They serve an important second purpose, by breaking up the text.

Use white space to make your text easier to read. A little extra space in the margins and between paragraphs can make a big difference.

Visuals can supplement your words, convey information, and add another dimension to your documents. Make sure your visuals are appropriate, accurate, and accessible.

8. Write actively (subject-verb-object) rather than passively.

Readers tire easily of the passive voice. Researchers recommend that only 10% of your writing use the passive voice. To write actively, avoid the verbs "to be" and "to have" and emphasize the subject-object-verb construction.

9. Avoid negative writing.

Accentuate the positive—particularly if you're dealing with negatives. Begin your memo or letter on a positive note. If you must present something negative, give the reasons: they usually help soften the blow.

10. Use the power of persuasion to influence readers.

Motivate your readers from the start by establishing common ground or giving them benefits. You may want to present a problem and offer a choice of solutions. End a persuasive message with an action close. Tell your readers what they should do.

Try these persuasion pointers:

- Know your readers—and constraints, factors keeping them from agreeing with you.
- Know what you can accomplish.
- Establish mutual goals/common ground.
- Be clear.
- Be familiar.

- Anticipate objections and dispel them.
- Use a problem/solution format.
- Give options or choices.
- Stress benefits for your readers.
- Ask for what you want.

These top ten tips will help you write more effectively in any business situation. Put these tips to work for you and write well!

Quick Reference to Writing Basics

Wise advice sometimes comes from opposite directions. Somebody may tell us, "Don't sweat the small stuff," while somebody else reminds us, "The devil is in the details." When it comes to writing, the truth lies somewhere in between those two extremes.

The essence of effective writing in business is to get the recipient of your message—whether it's a memo, an e-mail, a letter, or a report—to understand your message and, if your purpose is to persuade, to act upon that message. Put simply, you want your writing to work. From that perspective, such matters as style, vocabulary, grammar, spelling, punctuation, and layout may seem unimportant.

However, communication connects humans—and humans are often affected by style, vocabulary, grammar, spelling, punctuation, and layout. So, from that perspective, those matters may be very significant, since they may very well influence the effect of your writing.

There are hundreds of books out there that will promise to tell you everything you need to know about style, vocabulary, grammar, spelling, punctuation, and layout. But if you don't have the time, energy, or interest to study those matters in depth, we encourage you to take a few moments to read through this appendix—and to refer back to it whenever you write.

Style

The term "style" refers to the choices you make (consciously or unconsciously) with regard to several aspects of writing. Stylistic choices are not right or wrong, in principle, but only more effective and less effective, depending on your reader and the circumstances.

We'll focus here on the more important aspects and offer some recommendations. Bear in mind that stylistic choices should be appropriate to the type of writing—memo, e-mail, letter, report—and your purposes, and, of course, your readers.

In most business settings, you should write informally. That means that you should mostly use the vocabulary that you would use if you were conversing with your reader. But you should pay more attention to sentence structure than when you're talking with someone else. Opinion is divided over the use of contractions; the best advice is to use them if you know or believe they fit the circumstance of your communication from the perspective of your reader. The main point to remember is that we tend to write less effectively when we write to impress.

That brings us, quite naturally, to the question of reading level. As we've recommended, you should make your writing easy to read but not boring or simplistic.

Use the Gunning-Mueller Fog Index to make sure you're writing at a level that your reader will find comfortable. If in doubt, write at the reading level of the average 12-year-old. Sure, your readers are more literate than that, but they may not totally concentrate on your words.

Two aspects of style go together: abstract–concrete and general–specific.

It's always better to be as concrete as possible; abstractions tend to make writing less effective—particularly if you're writing to persuade. A good way to anchor your writing concretely is by providing examples. After explaining the details of a new profit-sharing plan, for example, you might tell how it would affect the average worker, using typical figures to illustrate. (Notice how we used an example to show what we meant by being more concrete.)

It's also usually more effective to be as specific as possible, because you then reduce the possibility of ambiguity. Why use the word "document" instead of "annual review" or "communication" to refer to a "memo"? In most situations, it's just more effective when we use specific words rather than general.

Sometimes our stylistic choices make our writing too specific. That's what happens when we use "businessman" instead of "business person," for example, or when we talk about finding "the right man for the job" when the "best man" could very well be a woman. It's just bad business to use words that make some people feel excluded.

Another matter of style is rhythm. Suppose we write in the style of primers. Then Dick can understand us. Jane can understand us, too. Even Sally can understand. If we use short sentences with a basic structure, Dick, Jane, and Sally can all understand us—but they're likely to stop reading out of boredom. Vary the structures and lengths of your sentences. And never let your sentences run on and on line after line—unless you're trying to lose your reader.

The last aspect of style we'll discuss here is punctuation. To some extent, that's a question of following the rules. But it's also a question of making choices—sometimes unconsciously. Let's look briefly at four examples of punctuation style:

- According to the rules, you can end a sentence with an exclamation point! But the previous sentence seems an inappropriate choice.
- You can use a dash to break up a sentence for emphasis but only in an—appropriate place. Not as we did just then.

- You can enclose occasional words, phrases, and even clauses in parentheses. But if (for whatever reason), you do this often (perhaps every sentence), you will usually (although not always) annoy your readers (at least after a while).
- You can use ellipses to trail your sentences off ... That may be useful after a list of items, examples, points, ... But don't use ellipses often ... Abuse of this punctuation endangers effective use ...

Some people use punctuation to impress their personal presence into their words. That's fine, as long as you don't overdo it.

In fact, that's the essence of style—making choices, but making them appropriately and wisely, to write most effectively.

Vocabulary

The words we use when we write are a question of choice. But sometimes those choices may be simply wrong for the context. The following list covers a variety of words that people commonly misuse. This is just a general list; you should add any words that you tend to use incorrectly.

accept, except—accept means to receive; except means to omit

affect, effect—affect is a verb meaning to influence; effect is usually a noun, but it can also be a verb meaning to bring about

all ready, already—all ready means that everyone is ready, already means previously

among, between—use between with two persons or things, use among with more than two

assure, ensure, insure—assure means to give confidence, ensure means to make certain, insure means to indemnify or safeguard

cite, site, sight—cite means to mention, site is a location, sight is a sense

complement, compliment—a complement completes, a compliment is an expression of praise
continual, continuous—continual means repeated often, continuous means without interruption
credible, creditable—credible means believable, creditable means deserving esteem
eminent, imminent—eminent means distinguished in a profession, imminent means threateningly near at hand
fewer, less—use fewer for countable objects and less for measurable quantities
imply, infer—when somebody puts something into the words, to suggest, that's imply; when somebody gets something out of the words, to conclude, that's infer
it's, its—it's is a contraction for "it is," its is the possessive form of the pronoun "it"
principle, principal—principle means law or truth, principal means main (adjective) or leader (noun)
there, their, they're—there is an adverb indicating location, their is the possessive form of the pronoun "they," they're is a contraction for "they are"
whose, who's—whose is a possessive form of the pronoun "who," who's is a contraction of "who is"
your, you're—your is the possessive form of the pronoun "you," you're is a contraction of "you are"

That's a long list of problem words—and you should make it longer as you add your "personal favorites." But if you keep it close to your desk and consult it often, you can tame those wild words and use them properly to make your writing more effective.

Here are some more words that many people use incorrectly:

crisis. If you've got more than one, it's crises. The same switch from –is (singular) to –es (plural) applies to analysis/analyses, hypothesis/hypotheses, and parenthesis/parentheses. It's all Greek!

criteria. This is the plural of criterion. Many people wrongly use it as if it were also singular. Another noun that works this way is phenomenon (singular)/phenomena (plural). That's really not complicated; it could be worse. Consider what happens to college graduates. A man is an alumnus, but two or more are alumni. A woman is an alumna, but it's alumnae for more than one. That's almost a lesson in Latin declensions!

e.g. This abbreviation of the Latin *exempli gratia* means "for the sake of an example." Use it wherever you could use "such as" to cite examples. Example: "We've invested in other businesses, e.g., a grocery store, a golf course, and two boutiques. Don't confuse e.g. with **i.e.**, the abbreviation of the Latin *id est*, which means "that is." Use i.e. whenever you could use "that is to say" or "I mean." Example: "Put the appropriate information, i.e., to, from, date, and subject, at the top of the memo."

former. Be careful when you use this word. If you write, "He was a former president," the past tense and "former" together mean that he's dead. Isn't he still "a former president"? You probably mean, "He was formerly president."

irregardless. Avoid this word; most experts consider it substandard. It's certainly redundant, with the prefix "ir-" and the suffix "-less." Use regardless.

revert back. Since revert means to return, the "back" is redundant: it's there in the prefix "re-." There are quite a few words that people unwittingly tend to reinforce unnecessarily, such as prospects—no need to add "future."

This list could go on for pages and yet not cover all the problem vocabulary. The main point to keep in mind is that language is a tool. In most circumstances it may be good enough to use whatever works, such as when a person might use a wrench to hammer a nail or use a screwdriver to chisel away a block of wood. The wrong tool may work well enough, but people may judge that person who uses the wrench as a hammer and the screwdriver as a chisel. You can use the wrong words or use words incorrectly and still communicate well enough, but you should never be content with just "whatever works."

Finally, one last warning about using jargon. You can use technical language with people who understand that language. You can even use technical language to impress people who act as if they understand. But don't use it with people who are unfamiliar with those terms, not if you want to write effectively. When you must use a specific technical term, make sure you define or explain it.

Grammar

When many of us hear the word "grammar," we cringe thinking back to sixth grade English when we were forced to stand in front of the class and conjugate irregular verbs.

To others, grammar connotes those picky points we don't need to worry about. But the use of correct grammar can make or break you professionally. Colleagues and customers expect you to use language correctly and not to make errors that educated people avoid.

Incorrect grammar can undermine your credibility—and it may even alienate your reader. English is a difficult language, but that's not a good excuse for some of the managerial writing we've read.

Language is like clothing. We could work naked and communicate with gestures and a few words here and there. But we dress at least well enough to adapt to social norms and we use language according to certain generally accepted standards. Many people in business also recognize the importance of dressing for success. But there are still a lot of managers who don't realize that it can be even more important to write for success, to be as careful with their words as they might be in choosing their clothes. Think about where your words are going, to represent you.

This section on grammar is very selective. We can't cover everything—and if we did, you wouldn't read it. But we hope that the short time you spend reading through the following observations will help you improve your grammar and write better because of it.

Adverbs and adjectives. We all know the difference between these two parts of speech. And we'd never write something like "You're working good here" or "She's doing a well job" (unless she's digging for water or oil).

But then we trip up occasionally on linking verbs. Those are verbs—such as appear, feel, look, seem, smell, sound, and taste—that can be linked with a modifier. Then they work like the verb be, which means the modifier has to be an adjective, not an adverb. Confusing? Maybe a few examples will help.

He felt bad (not *badly*) about sending the memo.
The project seemed good (not *well*).

The linking-verb problem occurs most often with the verbs feel and look, because sometimes it's correct to use the adverb well after these two verbs—depending on what you mean.

She feels good today. (In other words, she's *up*.)
She feels well today. (In other words, she's *healthy*.)

Agreement. Every verb must agree in number (singular or plural) with its subject. The same rule applies to possessive adjectives. That's the basic rule. However, certain violations of that rule are becoming more accepted for possessive adjectives.

correct: Each of the executives wants his or her own office.
accepted: Each of the executives wants their own office.
correct: The corporation published its annual report.
accepted: The corporation published their annual report.

Notice that we call such violations of the rule "more accepted." That doesn't mean by everybody or in all situations. It also doesn't mean it's acceptable. Be careful about casual violations of the agreement rule.

Parallel construction. Use a parallel structure for sentence elements joined by the coordinating conjunctions *and* and *or.*

incorrect: A student can learn typing, filing, and to write.
correct: A student can learn typing, filing, and writing.

Reflexive, unnecessary. There's a growing tendency to use reflexives as polite forms of the object pronouns. It's wrong, and it feels awkward. Consider the following uses:

incorrect: I'm doing fine. And yourself?
correct: I'm doing fine. And you?
incorrect: The authorities met with the managers and myself.
correct: The authorities met with the managers and me.

Use the reflexive when referring back to the subject ("You gave yourself a day off?!") or to reinforce a subject pronoun ("I would never do that myself").

Sentence. We all remember learning something in elementary school about what constitutes a sentence: a conjugated verb and a subject (expressed or implied). You probably feel fairly comfortable about using sentences. But maybe you're concerned about the dreaded fragment that your teacher would mark in red.

Don't worry about the occasional fragment. It can be quite effective. (We've used them here and there throughout this book—and our grammar check warned us about every one of them!) One word of advice about fragments: the shorter the better. Longer fragments can be confusing—and they suggest to your reader that you don't know what constitutes a sentence.

That vs. which. This issue confuses a lot of writers, but it's really not all that complicated—if you know the difference between necessary and unnecessary.

We insert clauses in a sentence to provide information. Sometimes the information is necessary for a proper understanding of what precedes the clause. Sometimes it's just additional information, nice but not necessary. That's simple enough, isn't it? (These clauses are usually called relative clauses, because the information they convey relates to an earlier word or idea.)

How do we attach these informational clauses to the rest of the sentence? We use that or which. Use *that* when the information is necessary to identify or specify. Use *which* when the information just tells us more. (The "that clause" is called a restrictive clause, because it limits ambiguity. The "which clause" is called a descriptive clause, because it describes. Nothing confusing there!)

One more thing. Since the descriptive (which) clause gives us unnecessary information, which we could omit, we set the clause off with commas. But since the restrictive (that) clause is necessary to the sentence, we treat it as important, and we don't use commas.

So, that's the difference between restrictive and descriptive relative clauses. Now, let's check them out in action.

> The project *that* we started yesterday may be costly.
> The Danes project, *which* we started yesterday, may be costly.

In the first example, we need the information in our relative clause to specify which project will be costly, assuming we have more than one. So, the clause is restrictive: we use that and no commas. In the second example, we understand which project (the Danes project), so the information in the relative clause is not vital. Our clause is descriptive, so we use which and set it off with commas.

That's easy! Now let's complicate it a little. Suppose we have more than one Danes project. Then the information in our clause becomes descriptive rather than restrictive—so we'd put, "The Danes project that we started yesterday may be costly."

It's really simple, once you understand the difference

between restrictive and descriptive. Then, you're just a quick thought away from getting your clauses right.

Verbs. English is confusing when it comes to verbs. Many people learn all about verb tenses in school, then forget about the rules.

Most regular verbs form their past tense and past participle by adding -d or -ed. (For example, "I worked" and "I have worked" or "We hoped" and "We would have hoped.") There are quite a few verbs that don't follow this pattern: the past tense and/or the past participle assume irregular forms. With some verbs, there's a difference between these two forms. We have to memorize these verbs and what happens with them. That's the way it is with the following verbs. (This list is definitely not complete!)

Verb	Past Tense	Past Participle
be	was, were	been
begin	began	begun
bring	brought	brought
choose	chose	chosen
go	went	gone
have	had	had
run	ran	run
speak	spoke	spoken
write	wrote	written

Many dictionaries contain lists of irregular verbs. If you can't find one in your favorite dictionary, consult a grammar reference or textbook and photocopy the list of verbs. Then keep it handy.

Does it really matter? Probably—and maybe a lot. Suppose you send out a letter in which you write, "I've spoke to Janet, so she brung her accountant, or the meeting could have ran on for hours." The recipient of your letter understands you, of course, but your misuse of verbs is likely to make a bad impression. It suggests that you lack education and your communications skills are poor. That judgment is probably unfair, but it's a natural reaction in many circles. So, take care with your verbs.

Who vs. whom. Use whom when the word serves as an object and who when it serves as a subject. Most people have little or no trouble using who as the subject correctly, at least in most instances. But the use of whom causes a lot of problems. For example, which sentence in each of the following pairs is correct?

Whom do you think will be promoted?
Who do you think will be promoted?

Whom did you expect to attend the meeting?
Who did you expect to attend the meeting?

The correct examples are the second and the third. There's an easy way to decide here between who and whom. Rework the sentence using he/him or she/her.

Which seems better, "I think her will be promoted" or "I think she will be promoted"? The second, so our correct example would be "Who do you think will be promoted?" Which seems better, "I expected him to attend the meeting" or "I expected he to attend the meeting"? The first, so our correct example would be "Whom did you expect to attend the meeting?"

Try the test on the following examples:

There's a bonus for whomever finishes on time.
There's a bonus for whoever finishes on time.

Where is whomever you chose to do this project?
Where is whoever you chose to do this project?

If you chose the second and third examples, read on. If not, you might want to read this explanation again.

One final word about who and whom: our attitude toward the rules is changing. Casual use allows us to say such things as "Who were you talking about?"—at least in conversation. Be careful when you write, especially in more formal situations. Then it would be safest to use the correct form, "About whom were you talking?"

(Notice how we avoided ending the sentence with a preposition. That's a similar case of a rule being ignored in casual, conversational use, but still usually respected in writing, at least more formally. The best advice for business writing would be to avoid ending a sentence with a preposition only if it's fairly easy to avoid. That may not be what your English teacher would advise, but we work in a more practical world than the classroom.)

You and I. Many people confuse the use of "I" and "me," particularly in certain set expressions—generally after *and* or *or*. (I suspect it's because so many teachers corrected so many kids who would say, "You and me are going to flunk English.")

How many times have you heard someone say, "Between you and I..."? It's incorrect usage, because between is a preposition, so we have to use the object pronoun me. People also say such things as "It's easy for you or I." Again, we should use me because for is a preposition.

One test that might help you decide is to reverse the pronouns connected by *and* or *or*. Most of us would never say, "Between I and you," since "Between me and you" sounds better—if we ignore for a moment the polite rule about never putting *I* or *me* before other pronouns. Likewise, we'd prefer "It's easy for me and you" over "It's easy for I and you." Our instinct kicks in when we reverse the pronouns, but then we ignore our instinct when we start with "you and." (We have fewer problems if we use pronouns other than *you*, because we distinguish between *he* and *him*, between *she* and *her*, between *we* and *us*.)

Bottom line. One final word about grammar. Don't trust the computer to check your writing. Sure, run the grammar check, if you've got one. It can't hurt—unless you trust it to catch every mistake and not to flag correct usage as incorrect. If you're weak on grammar, find someone who can check over the things you write—maybe a colleague, maybe an employee. Then pay attention to what he or she tells you about your writing.

Spelling

No, we're not going to overwhelm you with lists of words that are often misspelled. And we're also not going to remind you that the best way to improve your spelling is by reading prolifically. (That's true, but we can't advise extensive reading unless we can also suggest ways to get more time to do it!)

We're just going to urge you to *use the spell check function* in your word processing program—*but not rely* on it alone. Keep a dictionary on or around your desk. Proofread your work. Ask others to check over important letters and all reports. Finally, keep a list of all the words you tend to misspell. That should help you pay particular attention to your "favorite mistakes."

Punctuation

Apostrophe. The apostrophe usually marks omissions in contractions. It's also used to show possession—with one very important and often confusing exception. We don't use an apostrophe with the pronoun "it" to show possession, as in "the book and its readers." The apostrophe in "it's" means it's a contraction, such as we used in this sentence.

Colon. Use a colon after an independent clause to introduce a list, an example, or an explanation. Example: "We did not open an account with First National Bank for two reasons: they don't offer interest-bearing checking and their advertising is misleading." You should also use colons after the salutation of a business letter.

Comma. Most business writers get carried away and add commas every so often, as if at random. Use commas to indicate short pauses. The best way to tell if you need a comma is to read a sentence aloud to see if you pause. Just don't break up "natural units" such as a simple subject-verb unit (incorrect:

"The employee, resigned") or an article-noun unit (incorrect: "The, manager sent a memo").

There are two ways of punctuating items in a series. One style inserts a comma before the "and" or the "or"—"We manufacture widgets, gizmos, and doodads." Another style—the journalistic standard—omits that final comma—"We cut, style and color hair." It's no big deal. Do it either way—but be consistent.

Dash. Use a dash much like a comma—to pause or to set off a thought that is loosely connected to the sentence. As we advised above, under "Style," don't overuse dashes.

Ellipses. Remember those three dots that mean "and so on" or "whatever"? Example: "Please bring to the meeting any pertinent memos, letters, reports ..." Ellipses can be effective, but they tend to soften your words, to make them seem uncertain, tentative, indecisive, ... Use them sparingly.

Exclamation point. They show strong emotion—unless you overuse them. Like this!!! Then they're just funny or annoying!

Hyphen. These marks join two or more words that express a single concept when these words precede a noun, such as "on-the-job training" but not otherwise, as in "training on the job."

Parentheses. These marks are used to enclose material that is more loosely connected to the sentence. Parentheses tend to minimize the importance of the set-off element. You may enclose one or more complete sentences within parentheses. (If you do so, make sure to put the end punctuation inside the closing parenthesis—like this.)

Quotation marks. Use quotation marks to enclose titles of short works, like articles or speeches, or to enclose words taken from special vocabularies or used in a special sense. Use quotation marks to enclose quotes short enough to work into your text. (What's "short enough"? One guideline is three lines or fewer.) If you quote a longer passage, it's generally more effective to set it off as a separate block of text, indented on both sides, usually without quotation marks.

The rules about using punctuation with quotation marks are complicated:

- Commas and periods go inside closing quotation marks. *Example:* "The shipment will arrive," she said. "Don't worry."
- Semicolons and colons go outside closing quotation marks. *Example:* He told me, "You'll get a bonus"; consequently, I'm celebrating.
- Question marks and exclamation points go inside the closing quotation marks only when they are part of the matter being quoted. *Example:* She asked, "What do you know about that memo?" I couldn't believe she was asking about "that memo"!

Semicolons. Use a semicolon to separate two independent clauses (groups of words that could function as a sentence) that are not joined by a coordinating conjunction like "and" or "but." Example: "Be careful to turn off the printer; don't try to change the toner cartridge while the machine is still on." A semicolon signals a shorter stop than a period.

Use semicolons sparingly in business writing. Whenever possible, turn a semicolon into a period (just a slightly longer stop) and make two sentences. Example: "Be careful to turn off the printer. Don't try to change the toner cartridge while the machine is still on."

Bottom line. Punctuation marks separate and connect our words, telling readers about how the words interrelate. If you're familiar with the basic rules, you can use punctuation stylistically, as we discussed above under "Style." But make sure that you know what you're doing.

Layout

We've emphasized in several chapters the importance of formatting. We've recommended using heads and subheads to show your organization and break up the text. We've advised using white space to make your words stand out for easier reading. We've encouraged you to use visuals to supplement your words—as long as those visuals are appropriate, accurate, and accessible.

So we've already covered the most important aspects of layout. We'll simply make a few recommendations for typography.

Word processing and desktop publishing programs have made it relatively easy to embellish our words with dozens of features. The only thing such programs don't provide is good taste.

We can't provide taste in this book, either. But we offer the following layout suggestions:

- Use a font size that's easy to read without being annoying. Depending on your font type, an appropriate size would probably be 10, 11, or 12 points.
- Use serif type for body copy and sans serif type for headlines. Research shows that all those serifs—the small finish strokes on each letter—make bodies of text easier to read. Courier and Times are serif types. Arial and Futura are sans serif types.
- Use *italics* sparingly. It takes us 20% longer to read italic type than to read regular type. That also means that it takes more effort, so if you use italics extensively, you're likely to lose some readers and annoy others.
- Use **boldface** sparingly. Bold is easier to read than italics, but it can quickly become overwhelming if abused.
- Avoid using ALL CAPS, except for occasional emphasis and in report titles and main headings. It's a violation of established etiquette to use them in e-mail messages, WHERE ALL CAPS ARE THE EQUIVALENT OF SHOUTING AT YOUR READERS.
- Avoid underlining, except for occasional emphasis. Word processing programs have made it so easy to use italics and bold that we can leave underlining behind.

- Use heads and subheads to show your organization. Make them informative and interesting.

You don't need to know a lot about layout and typography to write more effectively. If you apply just these basic guidelines, you can make your texts easier to read and more interesting—without risking offending your readers.

Going Beyond This Book

As I mentioned earlier, there are hundreds of books out there to help you write better. Choose a guide that's practical for your business writing and avoid the books that only English teachers can read or put to use. (Tip: If you don't feel comfortable with the table of contents and the index, put it back on the shelf.)

But the best advice I can offer you is to find a comprehensive, modern style handbook to keep near your desk. Don't try to read it through. Keep it as a reference. And from time to time, ask a colleague whose use of language you trust to check over your memos, letters, and reports and mark problems. Then decide how you can resolve the problems. If you can't figure out what's wrong and how to fix it, consult your handbook.

Good luck with writing well!

Index

I

K

L

M

N

P

About the Author

Suzanne Sparks received a Ph.D. from Temple University in Mass Communication and is Accredited Public Relations specialist (APR) by the nation's largest public relations association, the Public Relations Society of America. She currently directs the graduate program at Rowan University and teaches both graduate and undergraduate public relations and advertising. She also serves as acting director of The Communication Institute, a consulting arm of the university.

Dr. Sparks consults with major, national organizations such as Wyeth-Ayerst Laboratories, Bell Atlantic Corporation, The American Baptist Convention, and Harron Communications Corporation to provide training and communication services. She recently completed two Schaum publications for McGraw-Hill—*Quick Guide to Business Writing* and *Quick Guide to Presentation Skills.*